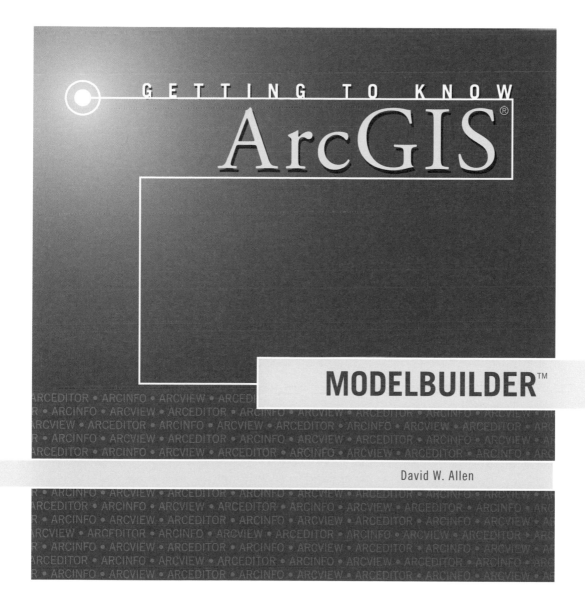

GETTING TO KNOW
ArcGIS®

MODELBUILDER™

David W. Allen

Esri Press
REDLANDS, CALIFORNIA

CONTENTS

PREFACE

In the past few releases of the ArcGIS product line, the ModelBuilder component has gained a lot of functionality. It has evolved from a simple tool that can be used to string a few geoprocessing commands together into a programming language that is controlled within a diagrammatic interface. In ArcGIS 10, ModelBuilder takes on more of a programming role by including the ability to iterate through datasets and features, to perform process looping and branching, and to incorporate scripts for complex calculations and decision making. ModelBuilder tools are available in ArcView, ArcEditor, ArcInfo, and ArcGIS Server. This makes the ArcGIS ModelBuilder application work not only as a stand-alone programming tool, but also as an interface for Python and VB scripts.

The challenge is learning how to use the ArcGIS ModelBuilder components effectively. While ArcGIS Desktop Help describes what the various ModelBuilder components do, this book goes a step further and illustrates how the components work together to create full processes. First, the book describes each component and its role in ModelBuilder, and then provides exercises to help the reader gain hands-on knowledge of how to string the components together effectively. This book will be helpful for the student who has limited exposure to GIS, the intermediate GIS user doing independent study, and the seasoned GIS professional seeking more advanced skills. It starts off simply and lays the basic foundation for building models but progresses to cover many of the more complex features of ModelBuilder. Readers should have a general knowledge of ArcMap and the geoprocessing tools used for GIS analysis but will be able to easily follow the steps in the exercises to create custom tools and automate processes. For the more advanced exercises that work with iterators and scripts, it would be helpful to have an idea of the syntax used in programming and know how to create an outline for an analysis project. Some prior knowledge of Python scripting would also be useful, although the basics are taught in this book.

A few years ago, I saw an ArcGIS ModelBuilder application used to model floods. The printed model display took up a whole wall and looked overwhelmingly impressive. At that point, I started learning all I could about the ModelBuilder process and how to use batch processing and iterations so that I could understand that model. Now when I see very large models, I can break them down into their components and be more impressed by how they carry out a function than by just their size alone. These years' worth of research and practice are condensed into this text so that you, too, will be able to understand the model-building process and build your own impressive, wall-size models. And if you're already building wall-size models, perhaps you'll learn a few tricks to using iterators and feedback variables that will keep your models nice and more succinct.

Start slowly and build a few models to customize a tool or streamline a process. Then start looking at an analysis project that you may be working on and see if you can create a Model-Builder application that performs the geoprocessing workflows for you. And you won't need the whole back wall to build your model!

Most of the data that is used here comes from a municipal environment. That way, the examples can be completed against this vector data without requiring the use of an extension, such as ArcGIS Spatial Analyst. Secondly, it's what I know and am most familiar with. But as long as you focus on the concepts and tools, the type of data you use won't make a big difference in terms of learning the ModelBuilder concepts and techniques. The ideas can be translated to any of the geoprocessing tools, use any of the ArcGIS extensions, and be applied to any data type.

Remember also that this is not a book on how to perform analysis; it's a book on converting an analysis project you design into a model. The analysis projects used in the exercises may seem simple, but they exist only to demonstrate the ModelBuilder techniques and not to demonstrate analysis theory. If you have a better solution for one of the projects described in an exercise, by all means, diagram that workflow and build the model. It will be good practice for model building in general.

To use this book, you must have a copy of ArcView 10, ArcEditor 10, or ArcInfo 10 (or a later version) running on your computer. The map documents used in the exercises will not open on earlier versions of the software. The DVD that comes with the book contains the exercise data you need, plus additional resources, but does not include the software itself. For instructions on installing the data, refer to appendix D.

David W. Allen, GISP
April 2011

ACKNOWLEDGMENTS

As the program coordinator for the GIS degree offerings at Tarrant County College in Arlington, Texas, I teach a lot of ArcGIS ModelBuilder techniques in my classes, but I don't have time to cover everything. Students routinely ask if there is a good reference book for ModelBuilder programming so they can study model building on their own. I've also given presentations at regional GIS meetings on programming with ModelBuilder and I get the same question. Well, for everyone who has asked when the book's coming, here it is.

I'd like to thank Esri Press and editor Carolyn Schatz, who walked me through the process to get this book published. The other great people at Esri Press who helped get this book off the ground include manager Peter Adams; acquisitions editors Judy Hawkins and Claudia Naber; cartographer Riley Peake, who carefully tested all the exercises; the production team; Esri Press marketing and distribution staff; and good friend Clint Brown, Esri director of software products, who always makes this technology seem so approachable.

Thanks also to Fire Chief Robert Isbell of the City of Euless, who helped me get the regional fire response call data for the Tarrant County fire department cooperative. And finally, thanks to the City of Euless administration, which allowed its rich GIS datasets to be used in the making of this book. While the data and processes are based in reality, all the scenarios are fictional and should not be associated with the City of Euless.

Chapter 1

Introducing model building

In the same vein as the other books in the Getting to Know series, *Getting to Know ArcGIS ModelBuilder* takes a step-by-step approach to modeling, using examples and providing user exercises to explain the concepts. Users with no knowledge of ArcGIS ModelBuilder will be able to easily start the process of building models, but even intermediate and advanced users, including intermediate students and GIS professionals, will be able to find a challenge in the later chapters.

The chapters are intended to provide a comprehensive outline of the capabilities of ModelBuilder, which are accessible across all license levels. The book starts with the simple basics of models and progresses to the more advanced features added in ArcGIS Desktop 10, ArcGIS Server 10, and ArcGIS.com, including iterators and several Model Only tools. The final chapters show how a model can be written to look and perform just like an out-of-the-box ArcGIS tool.

The purpose of models

Models represent a programming technique used in ArcGIS to string tools together to accomplish a task. The ModelBuilder interface in ArcGIS 10 is actually more of a flowchart schematic that guides the user to visually lay out the task than it is a coded programming language. Rather than seeing lines of coded text, users who are creating a model see boxes and ovals that are connected by flow arrows to represent the task. The flow represented by a model can be followed easily by tracing the arrows from one end of the model to the other to see the path. The tasks created are typically used in one of two ways: to model a particular project or to make a custom tool that automates a process.

A modeling project may involve investigating a process that occurs in reality, and then trying to simulate it in the model. You can run the model over and over to see how changes in the input parameters affect the result, which may let you analyze or make predictions about the reality you are modeling. You can explore many possibilities and examine the results very quickly. For instance, if you were looking for mountainous areas to grow coffee, you could build a model to look for land with a certain slope, average temperature, and annual rainfall. By changing the acceptable parameters for slope, temperature, or rainfall, you could investigate other possibilities for acceptable land. You could easily add more parameters that describe acceptable land, control their values, and investigate a wider range of results. Or you could use historical data to model the growth of a city. The results could be used to predict future development and plan the necessary infrastructure construction such as new utilities or streets.

Models can also be used to create a custom tool unique to your application. The tool could simplify or shorten a task and can be run many times. The tool can also be shared with others. As an example, you might want to write a model that would create a new feature class with a predetermined set of attributes and environment settings for new well sites. Or perhaps you might need a tool to check and adjust the extents of a feature class of wild bird sightings before merging it into a common dataset of local flora and fauna. Others working with you could also use the model as a shortcut to a longer process.

Characteristics of models and their components

A basic truth about modeling is that models cannot perform a task that you cannot perform manually with ArcGIS geoprocessing tools. All the processes created in a model are composed of components that already exist somewhere in the ArcGIS framework. The rule is, if it doesn't exist in the ArcGIS framework of tools, you cannot do it in ModelBuilder. In fact, one of the first steps to creating a model is to do the process manually and record the steps so that they can be duplicated in the model. Only after all the steps are clearly defined can a model be created to automate the process.

Models are constructed from a variety of components that establish the workflow. Model elements, including tools, variables, and connectors, are the building blocks of a model, and text labels are added to document the components and explain what the model does. These components are edited in the model window on the model canvas and saved in custom toolboxes in ArcToolbox.

The completed models can be run in a limited way from the model window, or they can be run from ArcToolbox. Models can also be given input variables and rich documentation, so that when they are run from ArcToolbox, they will have dialog boxes that make the models look and behave just like existing ArcGIS tools. In fact, many of the tools available to document your model are used by Esri programmers to document the standard toolsets available in ArcGIS.

The tools that make up a model are ArcGIS commands and can be system tools created by Esri programmers or custom scripts that users write. They can also include model tools derived from another complete model or specialized tools written with specialized ArcGIS extensions or third-party programming tools. Tools are shown in the model diagram as orange boxes and have one or more inputs as well as one or more outputs.

Tool type	Description
⚒	**Built-in** tool. These tools are built using ArcObjects and a compiled programming language like .NET.
⚙	**Model** tool. These tools are created using with ModelBuilder.
𝒮	**Script** tool. These tools are created using the Script tool wizard and they run a script file on disk, such as a Python file (`.py`), AML file (`.aml`), or executable (`.exe` or `.bat`).
⚞	**Specialized** tool. These tools are rare -- they are built by system developers and have their own unique user interface for using the tool. The ArcGIS Data Interoperability extension contains specialized tools.

Tool category	Description
System tools	System tools are those tools built and delivered by ESRI. They are installed by ArcGIS or any of its extension products. Almost all system tools are built-in tools, but you will also find system tools that are script or model tools. For example, the Spatial Statistics tools are all script tools, but since they are built and delivered by ESRI, they are considered system tools.
Custom tools	Custom tools are built by you. These are most often script or model tools, but they can be built-in tools as well. There are an infinite number of custom tools. You can download custom tools that other users have built by visiting the Model and Script tool gallery found on the Geoprocessing Resource Center. You can access the Geoprocessing Resource Center at http://resources.arcgis.com/

Models can include system and custom tools, designated by these icons.

The inputs and outputs for tools are called variables and can be either existing data, known as data variables and shown as blue ovals, or the output results of the tool, known as derived variables and shown as green ovals. Variables can also be values, which may be an input parameter or the result of a calculation that the tool performs.

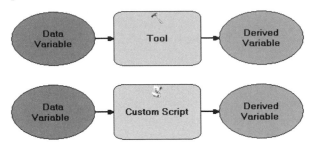

Variables contain the input and output data for tools.

The model components are then linked with one of four types of connectors, shown by different types of connector lines. This may range from the simple data connector that uses a project data variable or value variable as input for a tool, to the more complex feedback connector that can put a derived variable through several iterations of the same tool until a certain condition is met. The connector lines are shown as arrows and are designed to give the viewer a general idea of the sequence of the model's components and processes.

Models can also have text labels that help identify the various components. Each element may have a label that is tied to the box or oval it describes. When the element is moved, the text label goes along with it. It is also possible to add freestanding labels that are not tied to model elements. These may be titles or descriptions of certain processes that remain in the same place, even if the model elements they describe are moved.

Many users' first experience with models is to run a model that someone else wrote. The model may either perform a single task or act as a full application, but either way, using a model is a good way to share geoprocessing techniques with others. And running someone else's model is also a good way to see how models function and the various ways there are to run them. Before doing the exercises in this book, go to appendix D on installing the data, and load the exercise data DVD that came with this book.

Examining existing models

A friend has sent you a model that performs a single task. You will run it to see what types of processes can be contained in a model.

Before you begin the exercise, examine the steps needed to complete the task:

- View the model in the model window.
- Identify the model's components.
- Run the model.

Exercise 1a

1   Start ArcMap and open EX01A.mxd in the C:\ESRIPress\GTKModelbuilder\Maps folder. If you installed the book's exercise and sample data in a different folder from the default, go there and locate the Maps folder, and then open the map document.

2   Open the Catalog tab at the right of your map document by pausing the mouse over it. You may want to pin the Catalog window in place, since you will be using it frequently. Navigate to the GTKModelbuilder folder and find the Data folder. Expand it and find the Samples toolbox.

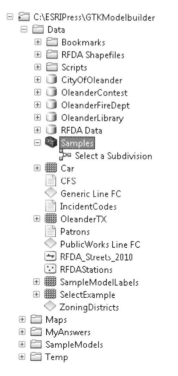

3 Find the Select a Subdivision model in the Samples toolbox. Right-click it, and then click Edit. An input variable, a process, and an output variable are displayed. Close the model window.

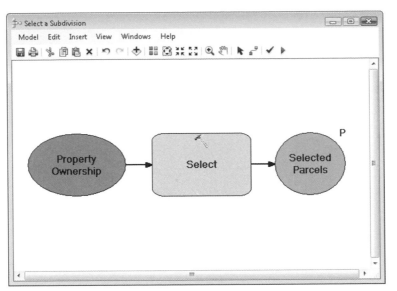

The model will perform a Select process on the Property Ownership feature class. The selection is preset to find all the features in the Running Bear subdivision. Later, you will see how to modify models to accept user input, which will make them more flexible.

4 Double-click the model in the Catalog window to run it. An input screen appears asking for the name of the output feature class. Click OK to accept the default.

The property selected by the model's process is added to your table of contents and symbolized (your colors may vary). This is a good example of automating a simple task with a model.

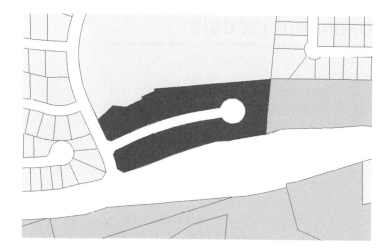

5 Close ArcMap.

What you've learned so far

- ◆ How to view the components of a model
- ◆ How a model's processes are run

Creating toolboxes and models

Models must be stored in a user's custom toolbox, which may be created in any folder or in a geodatabase. Users do not have the permission level to store models in an existing ArcGIS system toolbox, but new toolboxes are easily created. Once created, they can be shared among users, taking all the models they contain with them. This portability makes custom toolboxes, acting as shortcuts or custom tools, ideal for sharing models. More complex models used to study a particular project may have certain datasets associated with them that must also be shared with the toolbox in order for the model to function correctly in the new location.

Toolboxes stored in a folder can be shared with other users simply by copying the TBX files, while toolboxes stored in a geodatabase will require that the entire geodatabase be copied. There is also a My Toolboxes toolbox already created in the Toolboxes section of the Catalog tree. This is a shortcut to a user profile directory of the same name and is a convenient place to locate tools you expect to use across many sessions or in many map documents.

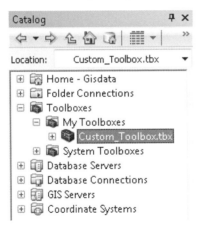

The Toolboxes folder can contain your custom toolboxes as well as the system toolboxes.

There are two ways to create a model. The quickest is to click the ModelBuilder button ![icon] on the ArcMap or ArcCatalog toolbar. An empty model window will open and the user needs to specify where the model will be stored. The other method is to right-click an existing toolbox, and then click New > Model, thus automatically specifying where the model will be saved. The default name is Model, but you should always rename your models to reflect their purpose. Imagine having three models called *Model 1*, *Model 2*, and *Model 3*. It would be difficult to know what tasks each model performs and whether you were using the correct one. If, however, they were named *Buffer Parcels*, *Union Parcels*, and *Select Parcels with Buffer*, it would be a lot easier to choose the correct model to perform the desired task. Once the model window is open, you can assemble the model components and test them on the model canvas.

The model canvas is used to assemble and connect model elements.

Dataset variables and tools can be dragged into this window to become part of the model. The next exercise deals with creating a new toolbox, creating an empty model, setting up environment parameters, and adding a few tools. Later exercises include working with other model components.

In the following exercise, you are the GIS programmer for the small city of Oleander, Texas, and you have identified tasks that you do repeatedly that might be good candidates for models. One in particular involves creating a new feature class with your standard spatial reference and a set of predetermined fields. The feature class will hold data about a single freeway accident, and a new one will be created for each reported incident.

Before you begin the exercise, examine the steps needed to complete the task:

- Set the location to store custom toolboxes.
- Create a new toolbox.
- Create a new model.
- Set the model parameters.
- Examine the Create Feature Class tool parameters.
- Save the model.

Exercise 1b

The first step in the process is to create the framework for storing the model.

1 Start ArcMap and open EX01B.mxd.

2   Open the Catalog tab at the right of your map document. Navigate to your MyAnswers folder where you loaded the exercise data DVD (e.g., C:\ESRIPress\GTKModelbuilder\ MyAnswers). Right-click your MyAnswers folder, and then click New > Toolbox.

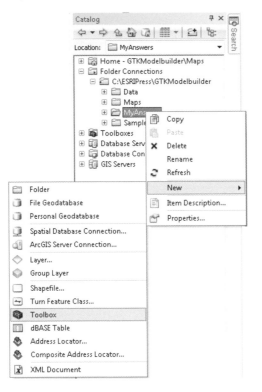

3   Name it **GTK Models.tbx** and press ENTER. This will hold your custom models.

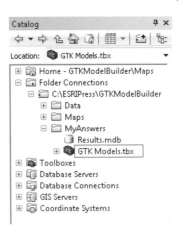

4   To create a new, empty model, right-click the GTK Models toolbox, and then click New > Model.

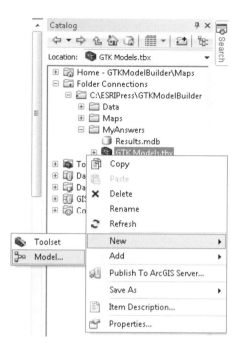

5   When the new model is created, the model window automatically opens.

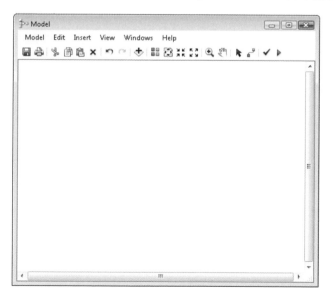

6 Next, you need to give the model a name, a label, and a description. On the ModelBuilder menu bar, click Model > Model Properties.

7 Type the name **CreateNewFC**, a label of **Create Oleander FC**, and a description of **Create a new feature class with the City of Oleander parameters and spatial reference preset.** Select the check box next to "Store relative path names," leaving the default check mark next to "Always run in foreground," and then click OK.

8 Save the changes to the model's name by clicking the Save button 🔲. Then close the model window.

This process created an empty model where the tools and variables can be dragged into place. However, before the model can be built, you need to investigate which tools are necessary to complete the task. The purpose of the model is to create a new feature class with a suitable spatial reference defined for the City of Oleander. First, you will search the ArcGIS tools to see which ones might be appropriate.

9 Click the Search tab at the right of the map document to open the Search window. Type the words **feature class** and click the Search button (Q).

10 Scroll through the results, looking for a tool that you think will do the job. There is a tool called Create Feature Class that looks like it will work.

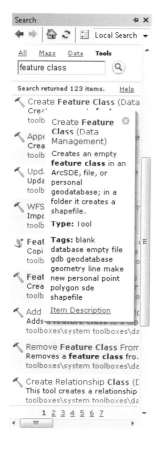

If you continue to look through the search results, you will notice a tool called Make Feature Layer and one near the end of the list called Make Table View. These two tools create a temporary file from the input data that does not persist outside the current session. They can be used inside a model to create temporary data files for analysis that do not need to be saved for later use. In this case, however, you will want to save the output file as a permanent feature class.

11 Click the Create Feature Class tool to open the Create Feature Class dialog box.

The first few inputs are for where to save the file, the name of the new feature class, and the geometry type. These are pretty typical inputs for creating a new feature class. The next input asks for a template feature class. It copies the data structure of an existing feature class and uses it on the new one. Field names, field types, default values, null flags, and any applied domains are copied. Since this model will be used to create many feature classes that have the same data structure, supplying a template is beneficial. Another optional input is the coordinate system. As in the normal feature class creation process, you can copy the spatial reference from an existing feature class to ensure that the reference is set correctly. With all this in place, it looks like this tool will work.

12 After reviewing the dialog box, close it. Close the model window and exit ArcMap.

This model involved creating the framework to store accident data for incidents that may occur over a length of time. When many accidents are recorded, the data could be used in an expanded model for analysis. You might want to predict where other accidents could occur based on the frequency of past occurrences, or you might want to find suitable locations for additional traffic safety devices.

What you've learned so far

+ How to create a new toolbox
+ How to create a new model, set its parameters, and save it
+ How to search for tools

Process states

Before you drag tools onto the model canvas, it is important to understand how they will react and how you can monitor their progress. Tools or variables in a model can be shown in one of three states: the Not Ready to Run state, the Ready to Run state, or the Has Been Run state. This is shown graphically so that the user can identify the overall state of the model.

A process in ModelBuilder is defined as a tool and the data variables it needs to run successfully. Processes in the Not Ready to Run state are shown as hollow wire-frame symbols with no color fill. This indicates that not all the required parameters of the process have been supplied. For instance, if the buffer distance has not been provided for the Buffer tool, the tool is not ready to run. Yet a parameter marked as optional will not keep an object in the Not Ready to Run state.

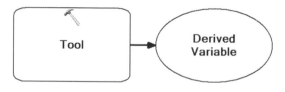

An incomplete model in the Not Ready to Run state

A process in the Ready to Run state has all the required parameters provided, either through user input or another variable. Note that although a process may be ready to run, there may still be optional parameters that you want to provide. For instance, the Create Feature tool listed the template feature class as optional. Not providing it would not keep the process in a Not Ready to Run state, but it would be of great benefit to the outcome of the process if it were provided.

A model in the Ready to Run state

A process in the Has Been Run state is shown with a drop shadow. This indicates that the process the tool is responsible for has been completed and its outputs are ready to be used by the next process. Processes can be run one at a time, so it may happen that only portions of a model are shown in the Has Been Run state. More commonly, though, the entire model is shown in the same state.

A model in the Has Been Run state

Adding model objects

Adding objects interactively to the model canvas is very easy. You can either use the Add Data or Tool button ✛ on the ModelBuilder toolbar, or simply drag layers or tools into the model from the table of contents in ArcMap, the Catalog or Search tab in ArcMap, or the Catalog tree in ArcCatalog. You can also drag or copy and paste all the datasets and tools used in any result from the Results window. For the exercises in this book, you will be dragging the objects onto the model canvas. This is a much more visual approach, and ModelBuilder is all about visual diagramming.

Another way to get objects into ModelBuilder is to set them as a property of a tool that has been added to the model. When the property is set, the ModelBuilder editor automatically adds it as a data variable to the model canvas and draws the connector line. The data in the current table of contents can be added, or data can be added from any data location.

You've identified the tool you'd like to use for creating the accident location feature classes and located it with the Search window. In the following exercise, you will add it to the model canvas and set its parameters.

Before you begin the exercise, examine the steps needed to complete the task:

• Start editing the model.
• Add the Create Feature Class tool to the model.
• Set the tool's parameters using the Import option.
• Identify the process states.

Exercise 1c

1 Start ArcMap and open EX01C.mxd in the C:\ESRIPress\GTKModelbuilder\Maps folder. Then navigate to your MyAnswers folder and locate the GTK Models toolbox.

2   Expand the GTK Models toolbox if necessary, right-click the Create Oleander FC model, and then click Edit.

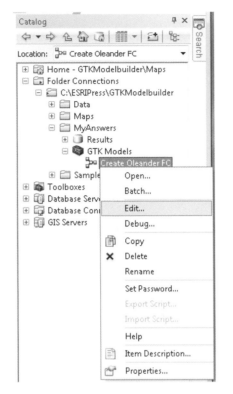

It is important to note that double-clicking a model or clicking Open on the model's context menu will run the model, not open it for editing.

The model window opens and is ready to accept tools and data. You will go back to the Create New Feature Class tool and drag it onto the model canvas.

3   Move the mouse over the Search tab to open the Search window. If your search results from exercise 1b are not still visible, search for Create Feature Class again. Click the Create Feature Class tool and hold the mouse button down. Then carefully drag the tool onto the model canvas and release it.

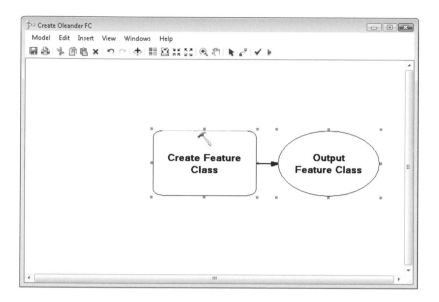

The tool appears in the Not Ready to Run state. The blue boxes around the tool indicate that both the tool's components are selected. These boxes can also be used to resize each component.

4 Double-click the Create Feature Class tool to view the Create Feature Class dialog box. Set the output feature class location to C:\ESRIPress\GTKModelbuilder\MyAnswers\ Results.gdb and type a new feature class name of **PD_1234** (a police incident number for the first accident). Next, set the geometry type to Point.

As discussed before, these are the only required parameters, but using a template file to add the required fields and setting the spatial reference will make the file immediately usable.

5 Set the template feature class to Accident Schema.

6 Now scroll down to the bottom of the dialog box and click the Browse button 📂 next to the coordinate system input box. Click Import and select the Accident Template feature class located at C:\ESRIPress\GTKModelbuilder\Data\CityOfOleander.mdb\ Accident Information.

7 Click Add and OK to set the coordinate system, and then OK to complete the setup for this tool. Click the Save button on the ModelBuilder toolbar to save the changes to the model.

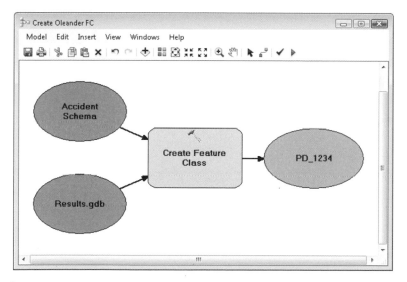

The two parameters that you set are now added as data variables. If your entire model diagram is not visible, click the Full Extent button on the ModelBuilder toolbar.

Can you identify the process state? The tool is in the Ready to Run state, since all its components have color but no drop shadow.

8 To review the parameters for the Create Feature Class tool, pause the mouse over the tool. When you have finished, close the model window.

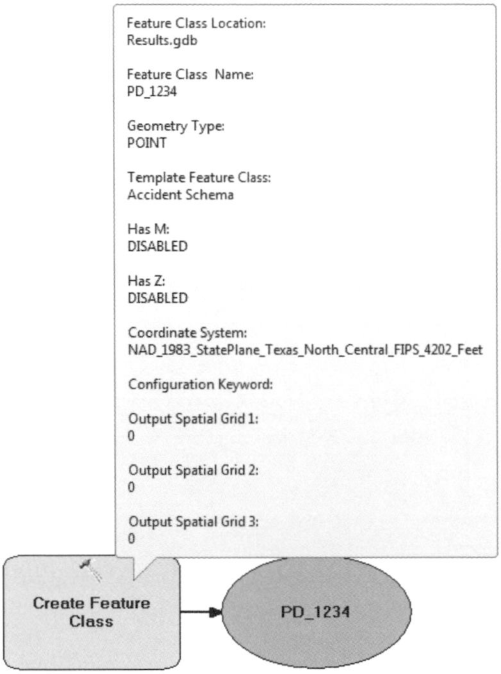

Feature Class Location:
Results.gdb

Feature Class Name:
PD_1234

Geometry Type:
POINT

Template Feature Class:
Accident Schema

Has M:
DISABLED

Has Z:
DISABLED

Coordinate System:
NAD_1983_StatePlane_Texas_North_Central_FIPS_4202_Feet

Configuration Keyword:

Output Spatial Grid 1:
0

Output Spatial Grid 2:
0

Output Spatial Grid 3:
0

Create Feature Class   PD_1234

What you've learned so far

- ◆ How to edit an existing model
- ◆ How to add components to a model and set their parameters
- ◆ How to identify a model's run state
- ◆ How to review tool parameters

Running a model

A model can have many components and a variety of tools, but they may not all be in the Ready to Run state at the same time. Some components may remain in the Not Ready to Run state until the entire model is completed, since they may be dependent on other components. However, you may still want to run some of the tools to check their functionality or to set up the model for the addition of another tool. There are several ways to run a model to accomplish this task.

The Model menu on the ModelBuilder menu bar has a command called Run Entire Model. This command will attempt to run all the tools in the model, even if they are not in the Ready to Run state. Even tools that are in the Has Been Run state will be run again.

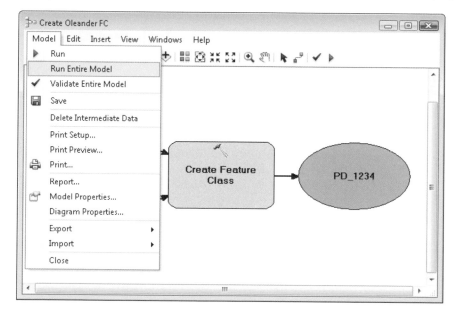

All the tools in your model will be run.

The same Model menu also has a command called Run. This command will run only the tools that are in the Ready to Run state. It will not try to run tools in the Not Ready to Run state, nor will it try to rerun tools in the Has Been Run state. The ModelBuilder toolbar has a Run button ▶ with the same function.

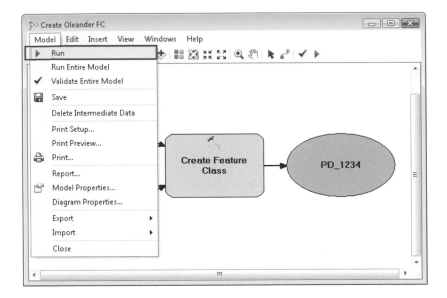

Only the tools in the Ready to Run state will be run.

The last way to run tools lets you run them one at a time. This may be necessary to set up variables for the next process, or you might just want to run a tool as a troubleshooting step to make sure it runs correctly. To do this, right-click a tool and then click Run on the context menu. Only the selected tool will be run. It is important to note that all the tools upstream from the selected tool that are in the Ready to Run state, and not in the Has Been Run state, will also be run.

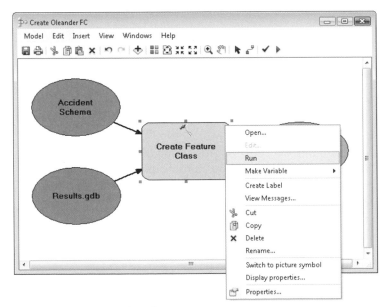

Tools can be run individually.

The model is complete, so the next step is to use one of the techniques described above to run the model. If everything is set up correctly, this will change the model's process state.

Exercise 1d

1 If you closed the map document after the last exercise, open EX01D.mxd and start editing the model Create Oleander FC. Be sure to click Edit and not Open. Once you have determined that the model window is displaying the model, use one of the methods described above to run the model. Watch closely as the model runs and you will see the tool currently being run turn red. With a complex model that contains many tools, you can follow the course of the processes visually. When the model has completed its run, it will go into the Has Been Run state.

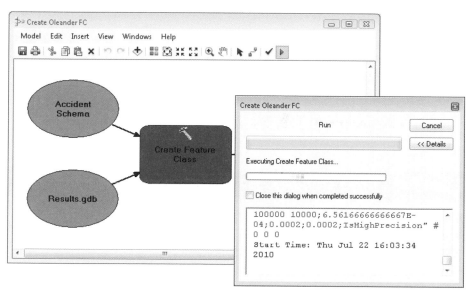

If you wanted to run this model again, you would need to click the Validate Entire Model button ✔ on the ModelBuilder toolbar to reset the model to the Ready to Run state.

2   Validate that the model has done what you asked by navigating to the C:\ESRIPress\ GTKModelbuilder\MyAnswers folder and examining the Results geodatabase.

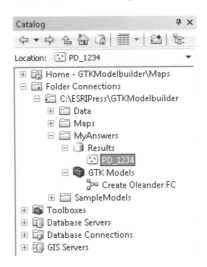

3   Complete the exercise by saving the model, closing the model window, and saving the map document. If you are not continuing, exit ArcMap.

What you've learned so far

- ◆ How to run a model using different techniques
- ◆ How to track a process as a model runs
- ◆ How to validate the results in the Catalog window

Navigating and laying out the model in the model window

As you draw a model, you will need to manipulate the model window to work with the model's components. This includes setting how the components are arranged on the canvas, zooming in, zooming out, and panning around the window. The ModelBuilder toolbar contains six navigation tools that control these actions. They are, from left to right, the Auto Layout tool, the Full Extent tool, the Fixed Zoom In tool, the Fixed Zoom Out tool, the manual Zoom In tool, and the Pan tool.

ModelBuilder navigation tools

The tools on the ModelBuilder toolbar have the following functions:

- The Auto Layout tool rearranges the elements in the model according to your layout settings.
- The Full Extent tool zooms to the boundaries of the model diagram.
- The Fixed Zoom In and Fixed Zoom Out tools zoom in or out on the view by constant increments.
- The Zoom In tool zooms to the area of a rectangle you draw.
- The Pan tool lets you scroll in the direction you want to view. Click the tool, and then drag the mouse in the model diagram.

These tools let you control your movements within the model window, but there are other tools and parameters that let you control how the model is displayed and how you view the different components. This can range from the way the window itself looks to the way the components of the model are drawn.

The look and feel of the model window can be altered by means of the Display Properties dialog box. A model's properties can be adjusted to draw a grid in the window, change the background color, link to a Web address, or even use an image as the background. When a component of the model is selected, the Display Properties dialog box controls how that individual component is displayed. The dialog box allows you to control the look of any tool or variable, and even use a custom image as the component's icon.

A model's look and feel is definable by the user.

When you placed a tool on the model canvas in exercise 1c, it was arranged to flow horizontally. The Diagram Properties dialog box lets you change this setting as well as others. The first of three areas you can control, shown on the General tab, deals with the model's automatic layout mode and grids. When the Auto Layout mode is set, the components are automatically arranged to display clearly and in the direction of process flow. You may also want to set up a grid to snap components and allow for clean manual placement of each element.

Grids may be used to control placement of model elements.

If you choose to use the automatic layout mode, the Layout tab holds global settings that determine how components are placed. One of the most important is Orientation, which establishes the direction for showing the model's flow. The default is horizontal, but you can choose any other direction. Another important setting is the Incremental Layout control. Setting this to Active means that only new components are arranged when the Auto Layout button ⊞ is clicked. This prevents the model layout from changing drastically as you add components. The last of the important settings are the Connection Routing settings, which control the spacing between components as well as whether the connector lines between elements run diagonally across the model or are staggered at 90-degree turns. The settings on this tab will take effect the next time the Auto Layout button is clicked, provided that you have set the model to operate in automatic layout mode.

Many of the layout settings are definable by the user.

Finally, you have control over the shapes and colors of the components via the Symbology tab. Two default styles are available, and by right-clicking a symbol you can control the color of the symbol and the color and font style of its text, including labels.

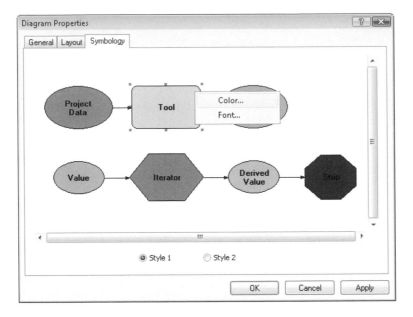

Users can control the styles, colors, and fonts of model elements.

All these settings are for visual effect only and do not affect how the model functions. Experiment with them to see how they make your models look and use them to set up a unique style for your models.

The model you created earlier was rather plain, using the default settings of the model window. In the following exercise, you will go back and change a few of the settings and examine the effects they have on your model's layout.

Before you begin the exercise, examine the steps needed to complete the task:

- Change the model's appearance with the layout tools.
- Create a picture symbol.
- Change the display properties.
- Change a symbol's color.

Exercise 1e

1   Open the map document EX01E, if necessary, and start editing the Create Oleander FC model on the model canvas. You should know how to do this by now. Click the Auto Layout button, and then click the Full Extent button to make sure all the model components are shown.

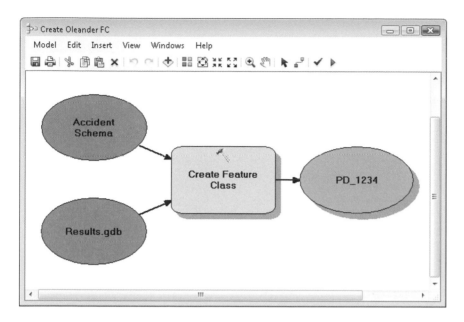

There's a car icon that might look better than a green oval for the PD_1234 output variable. Also, the Results data variable might look better as the Oleander city logo.

2 Right-click the PD_1234 oval, and then click "Switch to picture symbol."

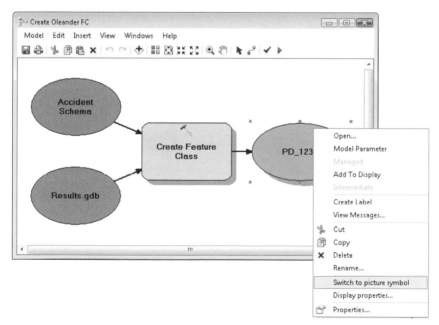

3 Change the file type to GIF and navigate to the Data folder. Click Car.gif, and then click Open.

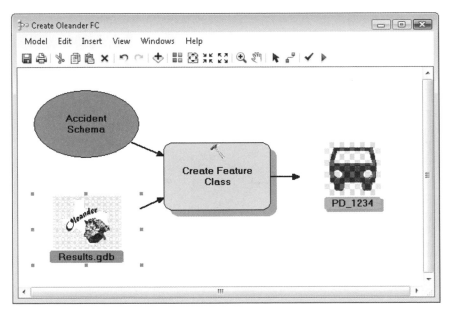

4 Repeat the same process on the Results.gdb oval and set the picture to OleanderTX.gif. This will add photos as icons for these components. You may need to click the Full Extent button to see them all.

5 Next, click the Model menu, and then click Diagram Properties.

6   Click the Layout tab. Set the orientation to Top to Bottom and the connection routing to Orthogonal Routing. Click Apply.

7   Next, click the Symbology tab. Right-click the box labeled Tool, click Color, and then click one of the purple colors. Click OK to save the color and OK again to close the Diagram Properties dialog box.

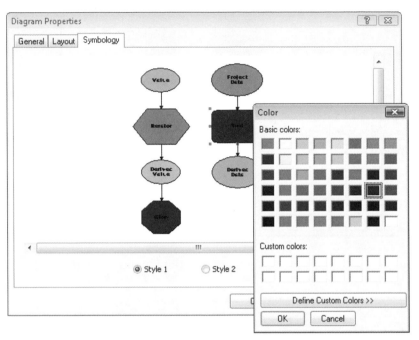

8   Finally, click the Auto Layout button and the Full Extent button to see the results.

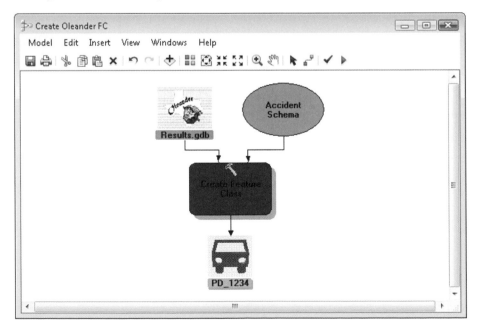

As discussed earlier, a lot more settings can be used to customize the look of a model. Feel free to try some of them out on your models.

9 Save the model and close the model window. If you are not continuing, close ArcMap.

What you've learned so far

- ◆ How to change the image used for model components
- ◆ How to change the properties of model components
- ◆ How to change the diagram properties of a model

Adding and connecting model components

Complex models are built by adding multiple tools to the model canvas, and then connecting them. Each required parameter of a tool must have an input variable, and it can create one or more output variables. These can then be used as input variables for other tools, creating a process flow to accomplish the objective of the model. Users can interactively link a data variable with a tool by using the Connect tool, or the connection can be created automatically if the data variable is assigned using the tool's properties dialog box.

The connector lines indicate the flow of the model's processes. In simpler models, the flow is linear, but as you'll see in following chapters, the flow can also branch and merge. A control process may be introduced to determine which of the branches runs. There may also be parallel paths in a model, with both sides required to run. A precondition can be set so that one side must run before the other. Unlike programming in scripts, this visual approach with models lets you easily follow the order of the processes and, moreover, allows you to easily explain the geoprocessing workflow to others.

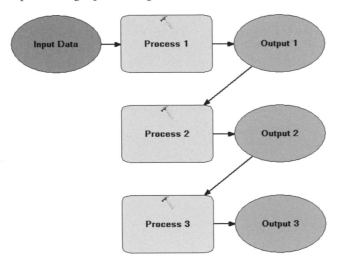

A simple linear model

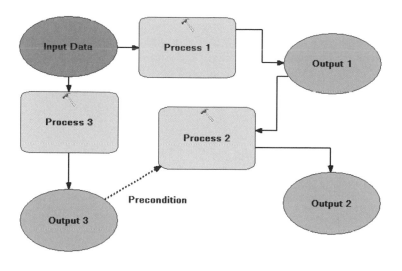

A more complex model with branches

The connector lines can be selected and their display properties altered. By default, connector lines are black with an arrow showing the direction of the process flow. But connector lines can be altered to change the style of the line, its thickness, the style of the arrowhead, and where the lines attach to the symbols. Connector lines can even be named and given a ToolTip.

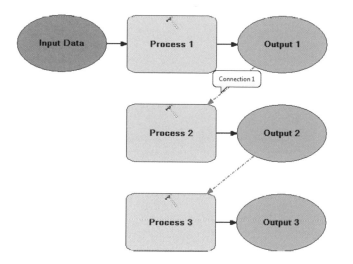

An altered connector line displaying a ToolTip

As you saw earlier, connector lines can also be altered to appear at right angles rather than diagonally. This is the orthogonal setting in Diagram Properties. Once the connector lines are drawn, they can be selected individually and redirected. The ones shown in the accompanying diagram were generated with the Auto Layout tool, and then the bottom line was selected and its nodes moved for clarity.

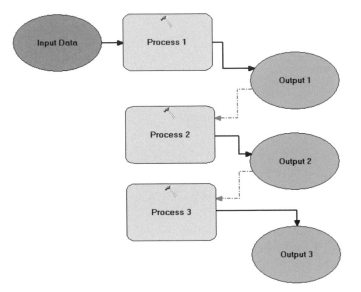

Connector lines constrained to right angles

Again, these settings don't affect the running of the model, but they will make your models look nicer and more organized, or they can be used as a means of visually grouping sets of processes.

Now that you've seen the basics of ModelBuilder, it's time to go through the full development cycle of building a model: first identify a process suitable for modeling, research the tools required, set up the environment for the model, drag the tools and data onto the model canvas, and then make the correct connections.

In the following exercise, the patrol sergeant in the police department is in charge of the scene for all accidents in the city. One task at the scene is to check the well-being of all the citizens within 500 feet of an accident site. This involves doing a door-to-door check of all residences and businesses to make sure that there have been no side effects from possible toxic fumes or chemical spills. Currently, the sergeant just sends some patrol officers up and down the street, hoping they check the right places. The patrol officers often duplicate their efforts, not knowing who is checking where, and they also miss houses periodically. The sergeant would rather have a checklist of all the property addresses within 500 feet of the accident scene that he could then divide among officers and know that they are staying on task, and not duplicating each other's efforts.

To put this scenario in GIS terms, in the exercise, you will select the point representing the accident location, buffer it 500 feet, use the resulting buffer polygon to select the property ownership polygons, and create a table with the list of property addresses. The table can then be printed and distributed to the patrol officers. This is a simple, linear process, and it sounds like a good candidate for a model. But before you start dragging tools around and connecting them, you should duplicate the effort manually and record the steps.

Before you begin the exercise, examine the steps needed to complete the task:

- Search for the Buffer tool and run it to test the process.
- Search for the Select Layer By Location tool and run it to test the process.
- Search for the Copy Rows tool and run it to test the process.
- Create a new model and name it.
- Set the ArcGIS geoprocessing parameters to aid in connecting model components.
- Add the three tools to the model and set their parameters.
- Use the Connect tool ᵷ to finish the model.
- Set the Add To Display property for the output.
- Test the model and save it.

Exercise 1f

1 Start ArcMap and open EX01F.mxd. A sample accident location is shown to use for the test. Use the Select Features tool to select it.

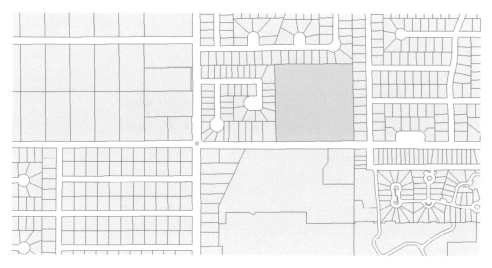

2 This selected point will need a 500-foot buffer. In the Search window, search for the word **buffer**. This will help you identify which tool may be required here.

3 The results show two possible tools, Buffer and Multiple Ring Buffer. The request is for a single 500-foot buffer, so the first tool is the one to use. Click it to see what information is required.

4 The tool needs an input dataset, which is the Accident Schema layer. Next, it needs an output feature class. You can navigate to your MyAnswers folder and have it create a new feature class in the Results geodatabase called **PD_1234_Buffer**. The last required parameter is the buffer distance, which you know is **500** feet. Enter these parameters and click OK to create the buffer.

5 The resulting buffer looks like it will work. You can see that it overlaps the features in the Property Ownership layer and you can do an overlay selection to get the common features.

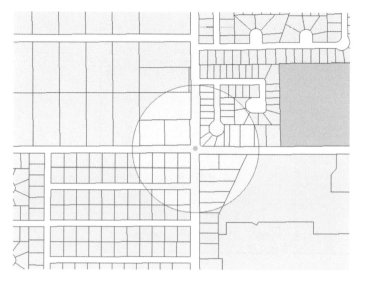

6 The next process is to do an overlay selection. This is normally done with the Select By Location tool under Selection on the ArcMap main menu bar. If you are not familiar with this tool, take a look at its dialog box. Now you need to see if there is a comparable tool for ModelBuilder to accomplish this type of selection. Do a tool search for **selection**.

7 The resulting list contains all the tools that perform some type of selection. Click Select Layer By Location and see if it compares to the manual version of the tool.

It is interesting to note that the Selection tool has an option to allow a buffer distance to be applied during the selection process. This would eliminate the need to create the buffer in steps 2 through 4 if the selection were the only important process. However, the patrol sergeant has asked that you retain proof of the selection buffer so that an exhibit map can be made for the file.

8   The Input Feature Layer is the Property Ownership file. These are the features to be selected. Next is the type of relationship to use for the selection. Intersection works fine for this process, so keep the default. But if you are curious, you can read up on the other choices in ArcGIS Desktop Help. Finally, the features to be selected are in the PD_1234_Buffer layer you created. Property Ownership features that intersect the PD_1234_Buffer layer will be selected. Click OK to run the process.

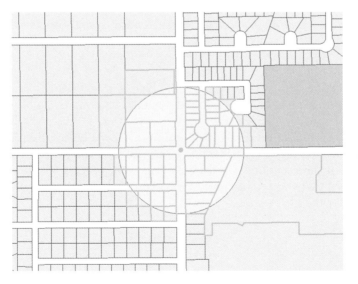

9   The result is that 53 features have been selected. These are the properties located within 500 feet of the accident.

10   In the manual process, you would simply open the attribute table and view or print only the selected features. But in the model, you will have to use a tool to copy the selected rows to a new table. Search the tools for **copy table**.

11   A lot of tools are returned by the search, and several of them could perform the desired task. The one you want to use to write the selected rows to a new table is called Copy Rows. Click it in the Search window to view its dialog box.

12 Set the input rows to Property Ownership. Then type the name of the output table as **PD_1234_Property** in your Results geodatabase. Click Save, and then OK to run the tool.

13 Click the List By Source button ⬇ on the table of contents and scroll down to find the new table. Right-click the table and open it to see the attributes. You'll see that there are fields that can be used to create a mailing list in a report. You could also include the land-use code so that the patrol officers doing the field work will know what type of structure they are going to.

The verification of the process was successful, and you discovered which tools will need
to go into the model:

- Buffer
- Select Layer By Location
- Copy Rows

Now you are ready to create the model, name it, and set up its operating environment.

14 In the Catalog window, navigate to the GTK Models toolbox you created earlier.
Right-click, and then click New > Model. Open Model Properties and type a name
of **AccidentBuffer**, a label of **Accident Buffer Analysis**, and a description of **Buffer the
selected accident and select property within the buffer.** Also select the check box to
store relative path names. Click Apply.

Accident Buffer Analysis Properties

General | Parameters | Environments | Help | Iteration

Name:

AccidentBuffer

Label:

Accident Buffer Analysis

Description:

Buffer the selected accident and select property within the
buffer.

Stylesheet:

☑ Store relative path names (instead of absolute paths)

☑ Always run in foreground

OK   Cancel   Apply

15 Next, in the Model Properties dialog box, click the Environments tab. You will want to
set up a default folder to contain the results of the model. Click the plus sign next to
Workspace to expand the list. Then select the check boxes for Current Workspace and
Scratch Workspace. Next, click Values.

16 Click the Workspace chevron and set both the current and scratch workspaces to the Results geodatabase in your MyAnswers folder. Click OK to make the changes. Finally, click OK to close the properties.

17 Next, you will set up an option that will make it easier to manually connect processes. On the ArcMap main menu bar, click Geoprocessing > Geoprocessing Options.

Geoprocessing	Customize	Windows
Buffer		
Clip		
Intersect		
Union		
Merge		
Dissolve		
Search For Tools		
ArcToolbox		
Environments...		
Results		
ModelBuilder		
Python		
Geoprocessing Resource Center		
Geoprocessing Options...		

18 In the ModelBuilder section, select the check box to display valid parameters when connecting elements. Then, in the Display/Temporary Data area, select the check box to add results of geoprocessing operations to the display. Click OK. Then save the model to preserve all the settings.

Geoprocessing Options

General
☐ Overwrite the outputs of geoprocessing operations
☑ Log geoprocessing operations to a log file

Background Processing
☐ Enable   Notification
Appear for how long (seconds)

Script Tool Editor/Debugger
Editor:
Debugger:

ModelBuilder
☑ When connecting elements, display valid parameters when more than one is available.

Results Management
Keep results younger than: 2 Weeks ▼

Display / Temporary Data
☑ Add results of geoprocessing operations to the display
☐ Results are temporary by default

OK   Cancel

With the "display valid parameters" option selected, each time you connect model components, you will be prompted for the correct parameter. Without this control over the connections, it is possible that they could link to the wrong parameters and cause errors in the model. The option to automatically add your geoprocessing results to the table of contents makes it easier to keep track of the layers and tables you may generate when you run your model outside the model window. Normally, such layers and tables are added by using the Add To Display setting on the context menu of an output variable, but this has no effect when running your model from the Catalog window.

You are ready to start putting data and tools into the model window and connecting them together. The first is the Buffer tool that will act on the Accident Schema layer.

19 Find the Buffer tool in the Search window and drag it onto the model canvas.

20 Double-click the Buffer tool in the model to open the tool's properties. Set the input features to Accident Schema, the output feature class to PD_Buffer in the Results geodatabase, and the distance value to 500. Click OK.

21 Notice that the Accident Schema data variable was added and connected to the buffer process automatically.

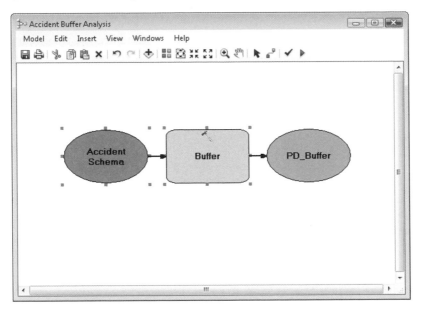

22 Locate the Select Layer By Location tool again and drag it onto the model canvas. You may need to zoom out a bit to see all the elements.

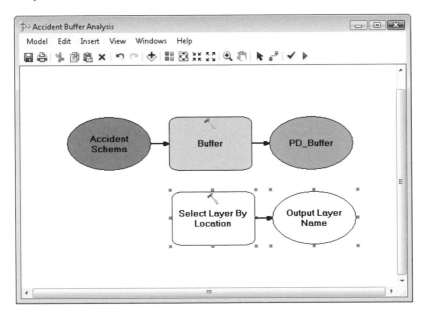

23 Click the Connect tool on the ModelBuilder toolbar. Click inside the PD_Buffer oval and then in the Select Layer By Location rectangle.

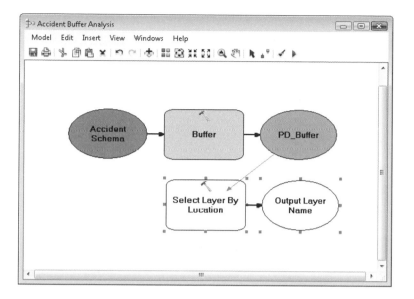

24 In the pop-up box, click Selecting Features. The Select Layer By Location tool will now use the buffer created in the first process to select features.

25 The tool is still in the Not Ready to Run state because it needs an input layer. Drag the Property Ownership layer from the table of contents onto the model canvas. Then use the Connect tool to connect it to the Select Layer By Location tool as the input feature layer.

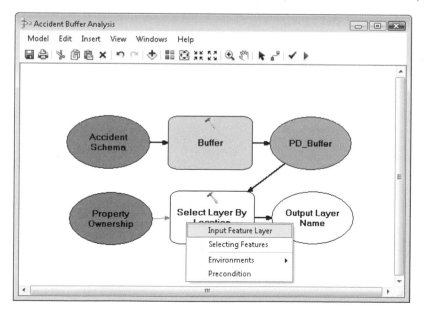

26 The selection process is now in the Ready to Run state. If you want to rearrange the model elements, use the Select tool ⬉ on the ModelBuilder toolbar to move them.

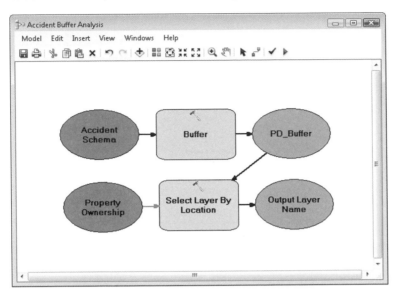

27 The last tool to add is the Copy Rows tool. Find it in the Search window and drag it onto the model canvas.

28 Use the Connect tool to connect the output from the selection tool to the Copy Rows tool as the input rows.

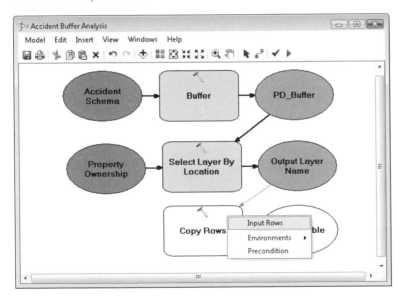

29 Double-click the Copy Rows tool to open its properties and set the output file to PD_ Notify in the Results geodatabase. Click OK.

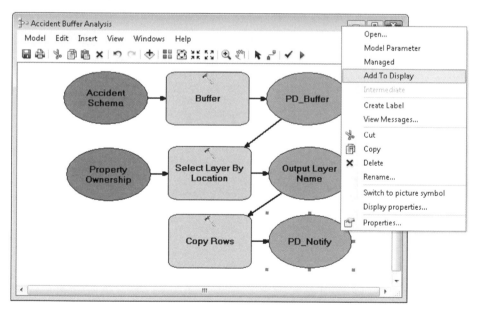

30 Since the output of the Copy Rows tool will produce the desired data, right-click the PD_Notify oval and select Add To Display. This will ensure that the output table is added to the map document.

31 The entire model is now in the Ready to Run state. Save the model.

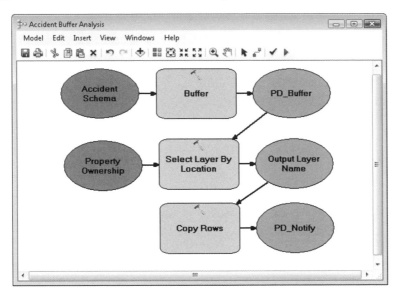

32 Move the model window to the side of your screen so that you can see the map document. In the table of contents, turn off the PD_1234_Buffer if it is still visible. Then select the accident shown at the intersection.

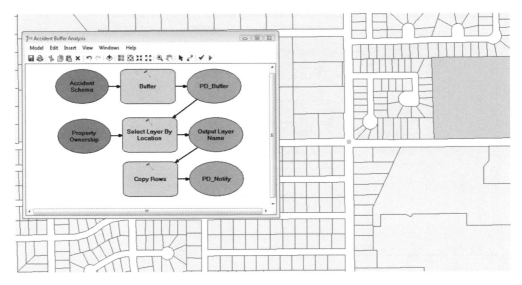

33 Click the Run button ▶ on the ModelBuilder toolbar and watch as the processes turn red as they are run. When the model has completed its run, all the processes will be in the Has Been Run state. Save the model and close the model window.

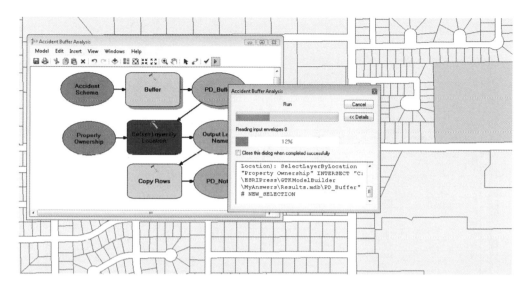

34 Click the List By Source button on the table of contents and scroll down until you find the PD_Notify table.

35 Right-click the table and open it to see the results. Notice that there are 53 features selected, the same number that the manual process produced. When you have finished looking, close the attribute table and save the map document.

Table						
PD_Notify						
	Prefix Direction	**House Number**	**Street Name**	**Street Type**	**Suffix Direction**	**Land Use Code**
		1617	MAXWELL	CT		A5
	N	1607	MAIN	ST		A1
		1619	MAXWELL	CT		A5
		200	EDINBURGH	DR		A5
		202	EDINBURGH	DR		A5
	E	106	ASH	LN		B1
	W	109	ASH	LN		A1
	W	107	ASH	LN		A1
	W	105	ASH	LN		A1
	W	103	ASH	LN		A1
	W	101	ASH	LN		A1
		1603	MAXWELL	CT		A5
		1605	MAXWELL	CT		A5
		1601	MAXWELL	CT		A5
		1607	MAXWELL	CT		A5
	E	109	ASH	LN		B2
	E	111	ASH	LN		B2
	E	201	ASH	LN		CITY

I◀ ◀   1 ▶ ▶I   (0 out of 53 Selected)

PD_Notify

The manual process of selecting property owners within 500 feet of an accident site is now automated in a model. The patrol sergeant can now select the accident feature and produce the address list without knowing anything about the underlying GIS processes.

If you like, open the model window again and try altering some of the diagram properties and properties of the elements to change things like the style of connector lines or the color and font of the variable symbols and their labels.

This model provided information that could be used to manage an emergency response situation or to predict where additional resources might be needed. The model could be expanded to create a driving path for the responders that would allow a complete search of affected properties from the generated list.

What you've learned so far

- ◆ How to search for and find tools in the Search window
- ◆ How to convert a manual process into a ModelBuilder process
- ◆ How to connect components in a model and set their properties
- ◆ How to set an output layer to display in the map window of ArcMap
- ◆ How to run a model and verify the results

Creating model printouts and reports

When completed models run successfully, it is often a good idea to print the final version for future reference. The schematic diagram that was used to create the model can be printed or exported to an image file, or a report of the model's elements can be generated.

Exporting the model diagram to a graphic is very simple and is accomplished from the Model menu. There are several output formats to choose from, including a JPEG or bitmap image, an Adobe PDF file, or a Windows Metafile. There are also controls that determine how much of the model is printed. For smaller models, perhaps the entire model can be exported. For larger models, the choices of Visible Window Only or Selected Objects Only can be used to export only a portion of the model. Other properties can be set to control the size and quality of the image.

An image of the model can be exported in many formats.

The results of the Export tool can be included in any documentation of the model or included with the model if it is shared with other users.

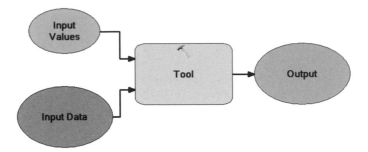

An image of the model can be used in other documents.

In addition to the graphic, a model report can be generated that will give the details of the model's parameters. The report has three sections. The first, called Model Report, shows the date and time that the report was generated. This can be matched against the latest edits of the model to make sure that the most current information is provided.

The second portion of the report, called Variables, details the variables used in the model. The data type and value of each variable are listed. Any messages that are generated about a variable when the model is run are also listed.

The final section of the report, called Processes, lists all the processes contained in the model. A process is a combination of input features, a tool, and output features, and each of these is documented in the report. Each tool name is shown, along with the location of the tool. This is helpful if tools are being used from custom toolboxes or if they are not loaded into ArcGIS by default. The parameters that have been set for each tool are shown. These include the variable name, whether it is input or output or is required for the tool to run, the variable data type, and the values that have been set.

The report finishes with a list of any messages that have been generated during the running of the model. The report can be shown on screen or exported to a file for archiving or sharing with other users.

A model report shows information about the model's tools and parameters.

The model you created in the last exercise is rather complex. The tool will be used mostly by others, so in the following exercise, you'll need to provide a diagram of what the tool accomplishes. In addition, you'll want to store the model report so that if any modifications are requested in the future, you'll have a record of all the model's elements.

Before you begin the exercise, examine the steps needed to complete the task:

- Export the model diagram to a graphic file.
- Generate a model report.

Exercise 1g

1   Start ArcMap and open EXO1G.mxd to access the Accident Buffer Analysis model you created in exercise 1f. Start editing the model on the model canvas.

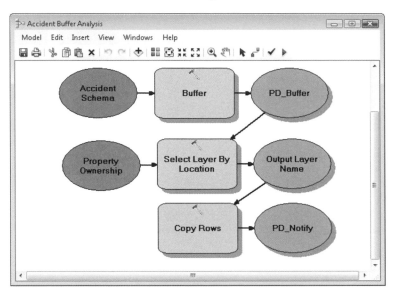

2   On the ModelBuilder menu bar, click Model > Export > To Graphic.

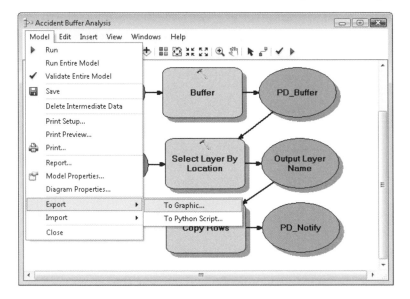

3   In the resulting dialog box, set the image type to JPEG. Set the output file name to Accident Notification Model in your MyAnswers folder. Review the other options in this dialog box for future use. When you have finished, click OK to create the image.

If you look in your MyAnswers folder, you will see the newly created graphic file. This file can be printed, sent to other users in an e-mail, included with reports, or used any number of other ways.

Next, you will generate a report about the model.

4 On the ModelBuilder menu bar, click Model > Report. Accept the default to view the report in a window and click OK.

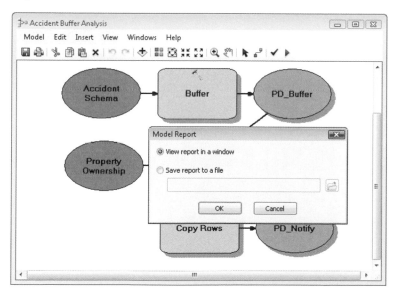

5 Click Expand/Collapse All in the upper-right corner of the report window. Review the first two sections titled Model Report and Variables. Note the values for all the model variables you created.

6 Scroll down the report and review the section titled Processes. Note the information provided about each tool used in the model.

As you work your way through the remaining chapters in this book, make graphics of the models you create and save the reports to your MyAnswers folder.

What you've learned so far

◆ How to create a graphic of a model diagram
◆ How to create a report that documents a model's components and parameters

Chapter 2

Setting up interactive models

One of the more powerful capabilities that you can add to a model is the ability to interact with the user. Models can prompt the user for input and output values or to set parameters for any of the model's processes. By giving a model the capacity to accept input, you can make it much more flexible, thus giving the user the ability to adapt it to a variety of situations.

Defining model parameters

Opening the properties of a tool and setting all its parameters is a quick way to add all the data required to put a model in the Ready to Run state. Models created this way can be run against the selected features, but if you run the model a second time, the output will overwrite the existing files. For example, a table created by the Copy Rows command at the end of a model process will always have the same name and will replace the existing file each time the model is run. Without defined parameters, a model would require no interaction from a user running it from the Catalog window or an ArcGIS toolbar. The only way to change any of the values used in the model's processes would be to edit the model in the model window and change the parameters of one of the processes in the window itself.

With no model parameters, the model does not accept user input.

Being able to prompt the user for the value of a variable is something all programming languages do, and the ArcGIS ModelBuilder application has a feature built in to give its variables this behavior. This is achieved by making a variable a model parameter. Any of the data, values, or output variables, shown as ovals in the model, can be made into a model parameter by right-clicking the variable, and then clicking Model Parameter. The variable is then tagged with a P in the diagram to show that it is a parameter, and when the model is run from the Catalog window or from an ArcGIS toolbar, a parameter input screen prompts the user for information before it continues.

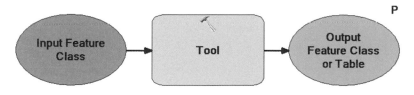

The output variable is set as a model parameter.

It is important to note that the user is not prompted to fill in model parameters if the model is run inside the model window. Once model parameters are set, the model *must* be run from the Catalog window or from an ArcGIS toolbar for the prompt screen to appear.

🖳 Images	⬜ ◻ ✖
Output Feature Class or Table	
C:\Users\David\Documents\ArcGIS\Default.gdb\Output_File	📂
OK Cancel Environments... Show Help >>	

The user is prompted for the name of the output file only when the model is run from the Catalog window or from an ArcGIS toolbar.

Within the tool properties, it is possible to make any of the parameters a model parameter, which means that the user will be prompted to enter a value for the parameter. That value is then used when the process runs. There are two steps involved in creating a model parameter: the first is to expose a tool parameter as a variable, and the second is to make that variable a model parameter.

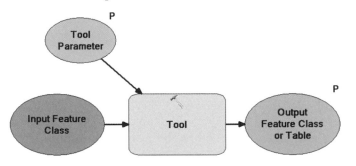

A tool parameter is set to accept user input.

Once model parameters are set, the input box displayed when a model runs looks very different. The user is prompted to fill in all the marked parameters before the model can continue, allowing for the outputs of multiple runs of the model to be saved.

This model has many user inputs defined.

Another thing that must be addressed when models are run from outside the model window is the capacity of the model to overwrite other data that does not need to be preserved. Some data, such as a temporary buffer or temporary table, may be needed as input for the next process but does not need to be saved for future use. It can be designated as intermediate data, and it is automatically deleted when the model has finished running. To make an output file serve as intermediate data, right-click it to expose the context menu, and then click Intermediate.

The check mark flags this variable as containing intermediate data.

Intermediate data does not persist past the completion of the model, but no data that is set as a model parameter can be flagged as intermediate. It must be saved to the disk when the model runs.

The model that you built for the police department accomplishes the prescribed task, but because the buffer distance and output file names must be defined during the editing process in the model window, it is difficult to use in the field. In the following exercise, there are no trained GIS operators on the scene to edit the parameters, and without this being done, the model will overwrite the results each time it is run. The patrol sergeant also tells you that most accidents get a 500-foot buffer, but if it is a freeway accident, or if there is a chemical spill involved, he may want to increase the size of the buffer. It would also be desirable to save the buffer for later verification after the sergeant returns to the office.

In the exercise, you'll change the model so that the user enters a name for the buffer that is created and a name for the final output table, and sets the buffer distance.

Before you begin the exercise, examine the steps needed to complete the task:

- Set two model variables as model parameters.
- Make the buffer distance value a variable and set it as a model parameter.

Exercise 2a

1   Start ArcMap and open EX02A.mxd. You may continue working with the model you created in chapter 1 or copy the Chapter 2a toolbox in the SampleModels folder to your MyAnswers folder. Start editing the Accident Buffer Analysis model.

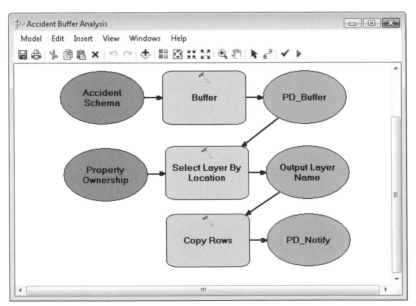

2 Right-click the output variable PD_Notify, and then click Model Parameter. A *P* is added next to the oval to show that it is now a model parameter.

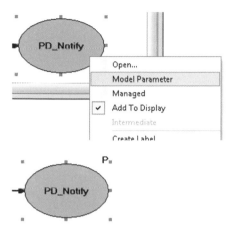

3 Repeat the process, making the variable PD_Buffer a model parameter as well.

4 Now you will make the buffer distance a model parameter. The first step is to make the tool parameter a variable. Right-click the Buffer tool, and then click Make Variable > From Parameter > Distance.

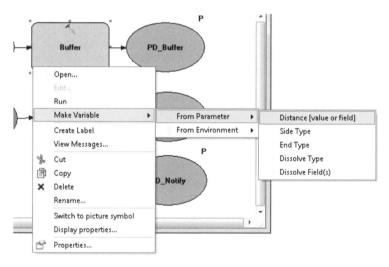

5   A new light-blue value variable oval is added and connected to the Buffer tool. Click it once so that it is the only selected object, and then drag it above the Buffer tool. Finally, right-click the new variable, and then click Model Parameter. A *P* is drawn next to it.

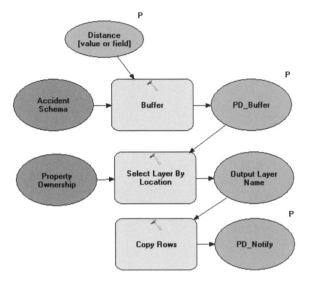

You may need to resize the distance variable to see its entire label and use the Full Extent tool to see the entire model diagram.

6   Save the model and close the model window. Now double-click the model in the Catalog window to run it and notice the input boxes that are presented. These correspond to the parameters you set.

7 Two of the parameters are marked as errors, because the file names already exist. For this test, assume that the incident number is PD_1122. Make the buffer file PD_1122_Buffer and the output table PD_1122_Notify. The errors should be cleared, so click OK to run the model.

8 Generate a graphic of the completed model, along with a model report. Then close ArcMap.

Now each time the model is run, the noted parameters can be set without a GIS specialist being needed on the scene to create this data. In addition, all the output files are saved so that further analysis can be done at a later date.

The addition of user interaction tools to this model makes it more flexible for handling a variety of situations. The model can be used to enter what-if scenarios to determine how the response might be expanded. The results can also be used to examine traffic routing around the incident and how that might change over time.

What you've learned so far

- How to set a variable as a model parameter and prompt the user for input values
- How to make a variable from a process parameter
- How to run models from the Catalog window

Establishing stand-alone variables and in-line variable substitution

The parameters of a process in a model can all be set as variables, whether they are standard data variables that hold the input and output features or value variables that hold parameters for the tools. These variables can also be set as model parameters so that the user is prompted to enter them before running the model. In addition to these variable types, there are stand-alone variables that are not derived from a model process. These variables can be any one of more than 110 data types, from feature classes to cartographic representations to sets of x,y coordinates.

Models can use variables from a large variety of data types.

After these stand-alone variables are created, they can also be set as model parameters. With the large variety of data types that can be used in variables, almost any degree of complexity can be added to your model. In addition, these variables can be used in any process within the model just by connecting them and specifying which parameter they will fill.

Some stand-alone variables, such as strings, Boolean values, or folders, may represent part of the value that is required for a parameter. For instance, the folder location string might need to be attached to the front of another value to create the entire output file name for a process. The value of the variable can be placed in the parameter input box by using variable substitution. This involves adding the variable name as a value bracketed by percent signs.

The string %FeatureClassName% is a variable whose value will be used as the file name.

The modifications made to the accident notification model let you set the names of the buffer feature class and the table of results, but the only real change from the default name was the addition of the incident number. If you were to set up a stand-alone variable to accept the incident number, you could substitute it in the parameter input box for both files and reduce the number of entries the user must make.

Before you begin the exercise, examine the steps needed to complete the task:

- Create a stand-alone variable and rename it.
- Make the new variable a model parameter.
- Use variable substitution to make the new variable part of the output file name.

Exercise 2b

1 Open EX02B.mxd in ArcMap. You may continue using the model you completed in exercise 2a or copy the Chapter 2b toolbox from the SampleModels folder to your MyAnswers folder. Start editing the Accident Buffer Analysis model. Clear the Model Parameter options for the PD_Buffer and PD_Notify variables by clearing those check boxes.

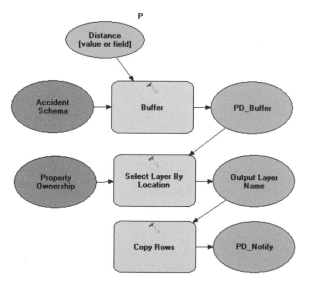

2 On the ModelBuilder menu bar, click Insert > Create Variable.

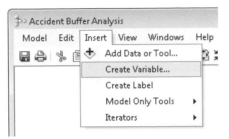

3 In the Create Variable dialog box, first note the check box for Multivalue. This feature allows the variable to hold multiple values. It is investigated in a later section. For now, scroll down and highlight String. Then click OK.

4 Right-click the new string variable and make it a model parameter. Then right-click it again and click Rename. Change its name to **Incident** and click OK.

5   Double-click the incident variable to open the Incident dialog box. Type the name **Incident Number** and click OK.

6   Now double-click the PD_Buffer variable to open the PD_Buffer dialog box. Add **%Incident%_** between PD_ and Buffer as shown. Click OK to save the changes.

7   Repeat the process and add **%Incident%_** to the output name of the PD_Notify variable.

The variables are no longer model parameters, so it would be a good idea to make sure they are not tagged as intermediate data. Notice also that there are no connector lines from the stand-alone variable to any of the model processes. The in-line substitution makes use of the values, but no connector lines are drawn. This also allows the variables to be used in more than one location throughout the model.

8 The model is ready to run. Save it and close the model window.

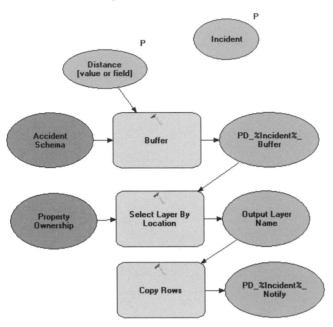

9 Make sure there is an accident location selected, and then double-click the Accident Buffer Analysis model in the Catalog window. Enter a buffer distance of **200** feet and the incident number of **2233**. Click OK.

As the model runs, the incident number you entered is substituted into the file names of both the buffer and the output table.

10 Add the PD_2233_Notify table to your table of contents and open it to view the records. When you have examined the results, close ArcMap.

What you've learned so far

- ◆ How to create a stand-alone variable and set its data type
- ◆ How to make a stand-alone variable a model parameter
- ◆ How to use variable substitution to place user input in a path or file name

Selecting features interactively

The previous model required that a feature be selected before running the model, because there is not an interactive selection tool in ModelBuilder. This process puts a burden on the user that they must understand what preparation is needed before running the model. The Selection tools used in ModelBuilder always require that existing features be selected either by their relationship to other existing features or through a query. A simple Select Features interactive tool is not available in ModelBuilder.

There is, however, a technique that can be used to select features using a temporary feature set created when the model is run. This involves creating a stand-alone variable with a data type of Feature Set, and then using a generic point, line, or polygon feature class as a template for the features. Once the model is run, the user is given a tool to draw these temporary features, which can then be used with a Select By Location process to select existing features from other layers. This type of interaction with the user helps to make the tools easier to understand and use without a lot of GIS knowledge or training.

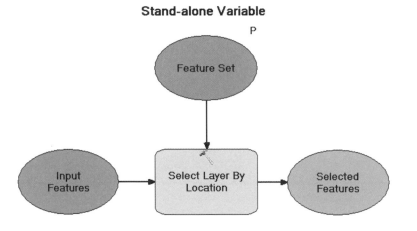

A feature set variable can be used to interactively select features.

The output is a set of selected features from the input feature class that can be connected to another process.

One interesting aspect of the feature set selection is that the attributes of the feature class that is used as a template are picked up by the feature set. When users input feature locations on the map, they can also enter attribute values. These values might be a text field that can be used for a query or a numeric field that can be used for a calculation. Any domains constructed and applied to the fields in the template file are copied and are active in the feature set. When the data variable is connected to a tool, the attribute values can be used as input and the existing domains are honored.

The feature set inherits attributes from the template file.

As you can see in the accompanying dialog box, there is also an option to add features from an existing source. This does not allow for user interaction with the map, but it can pull in data from another tool or process in the model. The advantage is that this dataset is temporary and does not persist beyond the completion of a model, so it can be an easy way to make a copy of a dataset and preserve the original data. This can apply to any type of dataset, including but not limited to feature classes, layer files, database tables, or Excel spreadsheets. Later, this book shows how to store and use these data types in temporary stand-alone variables.

In the following exercise, the public works department has come to you with another notification problem. The water crew will be doing some work tomorrow, and before they start, they want to put a notice on the doors of residents tonight letting them know that the street may be blocked off as well as the times the water service may be disconnected. This will involve selecting the road where the crew will be working and selecting all the properties that are adjacent to the road. A good way to do this would be to select the street centerline of the road where the work is to be performed, and then select all the features within 40 feet of the centerline. From the selected properties, you could then produce a checklist for the crews distributing the flyers. **Tip:** there are other methods to accomplish the same task, and you can explore these later on your own.

In the street centerline dataset, a street might be represented by dozens of small line segments that would be difficult to select manually before running the model. Instead, in the exercise, you will build in an interactive drawing tool for workers to use after the model has started, and then the model will use the one or two long lines they draw to do all the selections. This is done by creating a feature set variable, setting up its parameters, and making it a model parameter.

Before you begin the exercise, examine the steps needed to complete the task:

- Create a new model and name it.
- Create a stand-alone variable.
- Use a template file to set the new variable's symbology and data structure.
- Add the Select Layer By Location tool.
- Use the Connect tool and manual input to set the tool's parameters.
- Make the input data variable a model parameter.
- Add another Select Layer By Location tool and set its parameters.
- Add the Copy Rows tool and set its parameters.
- Set the output variable as a model parameter.
- Test the model and examine the results.

Exercise 2c

1   Start ArcMap and open EX02C.mxd.

2   Copy the Chapter 2c toolbox from the SampleModels toolbox to your MyAnswers folder. Create a new model in this toolbox and name it **PublicWorksNotify**, with a label and description as shown. Leave the check box to store relative path names selected.

3   On the ModelBuilder menu bar, click Insert > Create Variable. Set the data type to Feature Set.

4 Right-click the new data variable, and then click Properties. Then click the Data Type tab. The data type has already been set, but you will add a template file that adds predefined attributes and symbology.

5 Click the Browse button next to the "Import schema and symbology from" input box and navigate to the Data folder. Select the file Generic Line FC.lyr. Click Add and then OK to confirm the setting.

Look in: Data

Bookmarks	Generic Line FC.lyr
RFDA Shapefiles	IncidentCodes.txt
Scripts	Patrons.txt
CityOfOleander.mdb	PublicWorks Line FC.lyr
OleanderContest.mdb	RFDA_Streets_2010.shp
OleanderFireDept.mdb	RFDAStations.shp
OleanderLibrary.mdb	ZoningDistricts.lyr
RFDA Data.mdb	
CFS.txt	

Name: Generic Line FC.lyr   Add

Show of type: All filters listed.   Cancel

6 Next, find the Select Layer By Location tool in the Search window and drag it onto the model canvas. Connect the Feature Set variable to it for the selecting features.

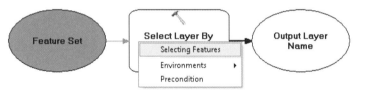

7 The tool is still in the Not Ready to Run state. Open it and set the input feature layer to Street_Centerlines. To ensure that the street centerline features are selected based on the lines drawn freehand, set the search distance to **50** feet. Click OK.

Select Layer By Location

Input Feature Layer
Street_Centerlines

Relationship (optional)
INTERSECT

Selecting Features (optional)
Feature Set

Search Distance (optional)
50   Feet

Selection type (optional)
NEW_SELECTION

OK   Cancel   Apply   Show Help >>

8 Finally, right-click the Feature Set variable and make it a model parameter. Rename and resize the variables if you like and arrange them on the model canvas so that they can all be seen. Save the model.

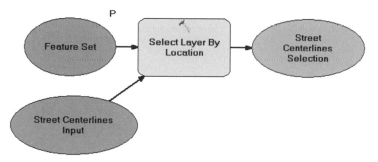

The first process is completed. The result will be a selected set of street centerlines. The next step is to use these selected centerlines to select the adjacent properties. Since no record needs to be kept of the buffer, it will be easier to use a search distance in the Select Layer By Location tool.

9 Drag another instance of the Select Layer By Location tool onto the model canvas. Set the input feature layer to Property Ownership, the results of the first selection for the selecting features, and the search distance to **40** feet. Click OK.

10 Rename the output variable **Property Ownership Selection**. The next process is now configured and in the Ready to Run state.

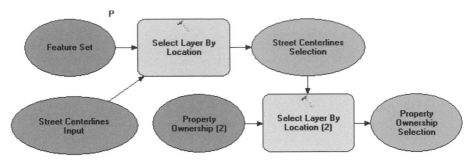

The final step is to use the Copy Rows tool as in earlier exercises to fill in a table with the selected records.

11 Find the Copy Rows tool and drag it onto the ModelBuilder canvas. Connect the Property Ownership Selection variable to it and set the name of the output table as a model parameter. When everything is in the Ready to Run state, save the model and close the model window.

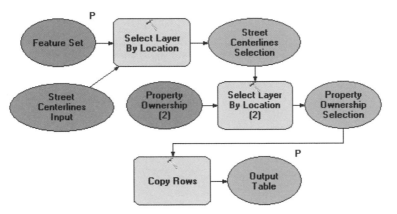

To test the model, you can use Kimble Drive. The procedure is to run the model, and then use the interactive Add Feature tool to draw a line down the middle of Kimble Drive. Then type a name for the output table, and the model should do the rest.

12 Double-click the Public Works Notify model in the Catalog window. In the Public Works Notify dialog box, click the Add Feature button .

13 In the map area, draw a line down the middle of Kimble Drive, double-clicking the last point to save the line. Notice that it is drawn with a preset symbol that was extracted from the symbology layer file template.

14 In the Output Table text box, navigate to your Results geodatabase and name the table **Kimble**. Click OK.

15 Refresh the map to see the selected parcels. Then open the results table by using the List By Source button on the table of contents window and review the records. When you have finished, close ArcMap.

Prefix Type	Prefix Direction	House Number	Street Name	Street Type	Suffix Direction
<Null>		301	KIMBLE	DR	
<Null>		303	KIMBLE	DR	
<Null>		305	KIMBLE	DR	
<Null>		307	KIMBLE	DR	
<Null>		309	KIMBLE	DR	
<Null>		1002	WINSTON	DR	
<Null>		902	WINSTON	DR	
<Null>		315	BRANCH BEND		
<Null>		904	WINSTON	DR	
<Null>		906	WINSTON	DR	
<Null>		302	KIMBLE	DR	
<Null>		300	KIMBLE	DR	
<Null>		304	KIMBLE	DR	
<Null>		306	KIMBLE	DR	
<Null>		308	KIMBLE	DR	
<Null>		310	KIMBLE	DR	
<Null>		312	KIMBLE	DR	
<Null>		908	WINSTON	DR	

(0 out of 33 Selected)

Kimble

A very simple user interface is presented, and the desired properties are selected. Test the model further, and try drawing several selection lines. You will see that multiple lines can be drawn and buffered, with the resulting table still containing the correct records.

Notice also that this was done without starting an edit session. This is possible because the only features you are editing are the lines in the newly created (and temporary) feature set. It is also important to note that even with an edit session active and snapping options set, the Add Feature tool does not allow you to snap to existing features.

This model allows the user to select other features around an area of interest. The resulting information lets you know how many people to notify and can be used to schedule street closures or to reroute traffic. A more complex version of this model could also include modeling the flow of water to the area to determine what effect removing a pipe from the network might have on water pressure.

What you've learned so far

◆   How to set up a selection interface in a model
◆   How to use an existing file as a template to set the symbology and file structure of a new layer

Using attributes in feature set selections

The Feature Set tool proved to be useful in exercise 2c, but it has another option that can extend its capabilities. The interface for selecting features also allows for the entry of attribute data. While the interface is temporary and is not saved once the model has finished running, it can be used while the model is active. Also, if a template file is used to control the symbology of the features in an output dataset, the attributes and their defaults and domains are inherited for use by the features. The dataset may be transferred to other permanent output files or used as part of the input parameters for one of the model's processes.

In the following exercise, the water crews have used this model to do their notifications, and now the public works director has asked if it can be modified for use by the road crews as well. Tomorrow, a road crew will be operating some heavy equipment repaving a street in another subdivision and wants to put notices on the doors of residents letting them know of a possible increase in noise during the day. They want to post flyers on the street being repaved, as well as on all the streets one block over in each direction. For this, you could use the interactive model to select the street centerlines, and then use buffers to select all the property within 400 feet, and from that, produce the checklist—the same process as before, except for the buffer size.

Making the buffer size a model parameter would let you change the buffer from 40 feet to 400 feet, depending on the job. You'd have to run the model twice, and you would produce two tables. A way to do this with a single run of the model would be to have the user enter a distance into one of the variables in the feature that they draw, and then use that value in the Buffer tool to produce two selection areas of different sizes.

In this model, you will need to construct buffers rather than use a search distance on the Select Layer By Location tool so that the attribute value can be used to size them. The search distance option on the Select Layer By Location tool only allows for a single distance, which would not work in this instance. You'll also be using a different LYR file for the input schema for the feature set.

Before you begin the exercise, examine the steps needed to complete the task:

- Create a new model and name it.
- Create a stand-alone variable and set the new variable's parameters.
- Make the new variable a model parameter.
- Add the Buffer tool.
- Use the Connect tool and manual input to set the tool's parameters.
- Add the Select Layer By Location tool and set its parameters.
- Add the Copy Rows tool and set its parameters.
- Set the output variable as a model parameter.
- Test the model and examine the results.

Exercise 2d

1 Start ArcMap and open EX02D.mxd. Copy the Chapter 2d toolbox from the SampleModels folder to your MyAnswers folder. Create a new model called PWDoubleNotify in this toolbox and add an appropriate label and description.

2 Begin editing the new model in the model window. Click Insert > Create Variable to add a stand-alone variable and set its data type to Feature Set. Open its properties and set the import schema and symbology layer on the Data Type tab to PublicWorks Line FC.lyr from the Data folder. Click Add to save the schema, and then OK to close the Feature Set Properties dialog box.

3 Right-click the new variable and make it a model parameter. Save the model.

4 Find the Buffer tool and drag it onto the model canvas. Connect the feature set to the Buffer tool as the input features. Notice that the tool is still in the Not Ready to Run state.

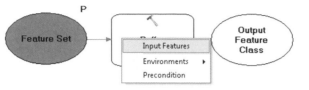

5 Double-click the Buffer tool to access the Buffer dialog box. Set the output feature class to C:\ESRIPress\GTKModelbuilder\MyAnswers\Results.gdb\PW_Buffer.

6 Under the distance setting, select Field, and then select Num_Integer. Click OK to save these parameters and close the dialog box.

Buffer
Input Features
Feature Set
Output Feature Class
C:\ESRIPress\GTKModelbuilder\MyAnswers\Results.gdb\PW_Buffer
Distance [value or field]
○ Linear unit
_____ Feet
● Field
Num_Integer
Side Type (optional)
FULL
End Type (optional)
ROUND
Dissolve Type (optional)
NONE
Dissolve Field(s) (optional)
☐ OBJECTID
☐ Num_Integer
☐ Num_Float
OK   Cancel   Apply   Show Help >>

The field Num_Integer has a coded values domain that will restrict its values to 40 or 400, which the feature set inherited from the template file. It also has symbology designed for these two values. If more buffer values are desired at a later date, they can be added to the field's domain and new symbology can be set for them.

7 The next process is set up the same way as in the previous exercise. Drag the Select Layer By Location tool onto the model canvas and use the Property Ownership layer as the input feature layer and the PW_Buffer layer for the selecting features. Note that there is no search distance, as that was handled by the Buffer tool.

Select Layer By Location

Input Feature Layer

Property Ownership

Relationship (optional)

INTERSECT

Selecting Features (optional)

PW_Buffer

Search Distance (optional)

Feet

Selection type (optional)

NEW_SELECTION

OK   Cancel   Apply   Show Help >>

8 Finally, drag the Copy Rows tool onto the model canvas. Use the output of the Select Layer By Location tool as the input layer, and make the output a model parameter. Change the variable names or sizes as necessary and arrange the model so that all the elements can be seen. Save the model and close the model window.

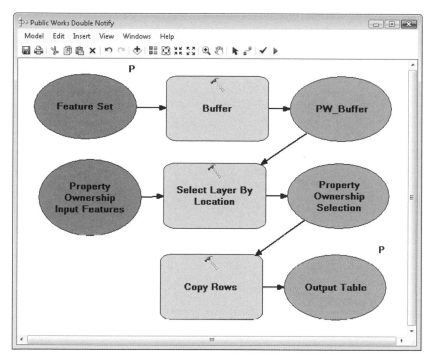

The only thing that remains is to test the model. For the test, mark Erica Lane and Amy Way as being repaved (a 400-foot buffer) and Carol Way as having water repairs done (a 40-foot buffer).

9 Double-click the Public Works Double Notify model to run it. As in the last exercise, draw a line along the center of Erica Lane, double-clicking it to end the feature. In the Public Works Double Notify dialog box, click the cell under Num_Integer and select 400 feet. Notice the change in the line's symbology.

10 Next, draw a line along the center of Carol Way and set the value for Num_Integer to 40 feet. Again, note the change in the line's symbology.

11 Finally, draw a line along the center of Amy Way and set the Num_Integer value to 400 feet. Change the name of the output table to **PW_Tuesday** in your Results geodatabase and click OK to finish running the model.

12 When the model has completed its run, open the table PW_Tuesday to see the results. When you have finished examining the table, close it and exit ArcMap.

House Number	Street Name	Street Type	Suffix Direction	Land Use Code	Plat Status	SHAPE_Length	SHAPE
702	BRENDA	LN		A5	1	336.637756	6813
705	BRENDA	LN		A5	1	338.795218	6547
704	BRENDA	LN		A5	1	335.091285	6989
707	BRENDA	LN		A5	1	409.901138	9574
600	CAROL	WAY		A5	1	336.199288	6501
602	CAROL	WAY		A5	1	340.893629	6638
604	CAROL	WAY		A5	1	356.567355	7577
601	CAROL	WAY		A5	1	353.41537	730

By adding the capability to define the needed buffers, you have given the model much more functionality. The streets and properties produced in the output list are valuable information that could be used to determine an overall plan of action for several months' worth of work. This could help determine street closures in advance and allow for a more comprehensive approach to keeping vital detour routes open.

What you've learned so far

- ◆ How to create new features in a model and use them for additional selections
- ◆ How to accept attribute information from user input

Symbolizing output

When using a stand-alone variable to accept input from the user, the user could provide a file as an input schema that controls the fields and symbology of the variable. As features are drawn, the fields would be populated and the symbols set.

It is possible to use a similar process to control the symbology of the output variables in a model as well. This is done by first creating the desired symbology in ArcMap. The symbology can be a simple set of unique features or a more detailed classification based on a numeric value. Once the symbology is set, a layer file can be created to store the settings. Remember that a layer file (.lyr) stores symbol settings but does not store any data. When the symbology layer file is specified as the template for the model's output, the settings are used to symbolize the output features.

The Output Feature Class Properties dialog box for the output variable includes a tab for layer symbology. The premade layer file is selected and stored for use when the output data is created.

The symbology from another layer is imported into the output file.

Any of the variables used in a model have this option. By using premade symbology, a lot of time can be saved in creating the output map. This technique is used in a later exercise.

You learned in this chapter that any variable or tool parameter can be set as a model parameter, which presents a prompt to the user for input when running the model. In one exercise, you learned that stand-alone variables can be created and used in in-line variable substitution, and even used to control interaction with the map. Finally, you learned how to control the symbology of not only the input variables, but also the output variables. Later, this book shows how to control the wording of prompts for the model parameters and how you can supply context-sensitive Help for your new tools.

1

2

Chapter 3

Establishing flow of control

If the ArcGIS ModelBuilder application is to function as a visual programming tool, it must be able to do the things that other programming languages do. One of these things is to manage variables and user input, which was demonstrated in the previous chapter, while another is to control the order in which processes are evaluated and run. This includes holding a process until another one is completed, running a process multiple times before moving on, and running (or not running) a process based on the status of another process or feature. Certain decision-making steps can be added to a model to establish these controls.

Defining preconditions

The models shown earlier were linear in nature; the processes followed a single line of control. Often, the order of processes may not be linear, such as having one process add a field to a table and then another process calculate a value into that field. If the second process runs first, there will not be a place to store the calculated value, and the model will crash.

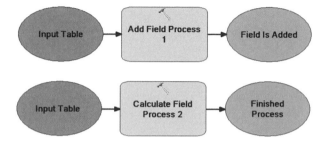

This model has no controlling precondition set.

In the accompanying example, the processes could run in any order. To prevent this, the first process is made a precondition of the second process. In other words, the second process cannot run until the first process is completed. This precondition can be set in one of two ways. The first is to open the tool properties of the second process and use the check boxes on the Preconditions tab. All the selected processes will become preconditions of this process.

The Add Field process must run first.

The second method of setting a precondition is to use the Connect tool and specify the connection as a precondition. This has the same effect as using the tool properties dialog box.

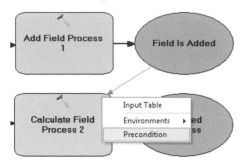

The connector's context menu can be used to set a precondition.

When a precondition is set, the connector line is shown as a dashed line. This indicates that the connected variable is not used as a parameter for the connected tool but the process that creates this variable must be completed before the next process can run.

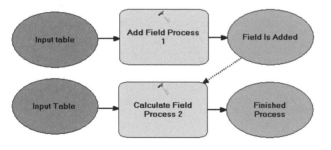

A precondition is shown with a dashed connector line.

Adding preconditions provides more control over how a model runs. Adding controls allows for more process components to run in a single model without making the model crash.

In exercise 1a, you created a model that would make a new feature class for each accident investigation. In the following exercise, the police detectives have decided after using the model a few times that they would like a new feature dataset created for each new feature class. Then the feature class can be created in the new feature dataset. As more analysis is done for each case, additional related data can be stored in the same feature dataset to avoid confusion with other cases.

In the exercise, you will add a stand-alone variable as a model parameter to prompt the user for the name of the geodatabase where the feature dataset will be created. Then you'll add a second stand-alone variable to prompt the user for the name of the feature dataset. Next, you'll add a final stand-alone variable to ask for the feature class name. Finally, you'll add the Create Feature Dataset tool to use the input from the user and the Create Feature Class tool to use the last user input to create the feature class. In order to ensure that all of this runs in the correct order, you'll make the Create Feature Class tool a precondition of the Create Feature Dataset tool and make all the user input variables preconditions of the Create Feature Class tool.

Before you begin the exercise, examine the steps needed to complete the task:

- Start editing the model.
- Create a stand-alone variable and set its parameters.
- Make the new variable a model parameter.
- Create two additional stand-alone variables and set their parameters.
- Make the two new variables model parameters.
- Set the parameters for the Create Feature Dataset tool by using variable substitution.
- Use the Import option to set the coordinate system for the new feature dataset.
- Set the parameters for the Create Feature Class tool by using variable substitution for the feature class name.
- Set a precondition for the Create Feature Class tool.
- Save and run the model to test it.

Exercise 3a

1 Start ArcMap and open EX03A.mxd. Copy the toolbox called Chapter 3a from the SampleModels folder to your MyAnswers folder.

2 Start editing the Create Oleander FC model. You'll notice that the two tools necessary for this model have already been added.

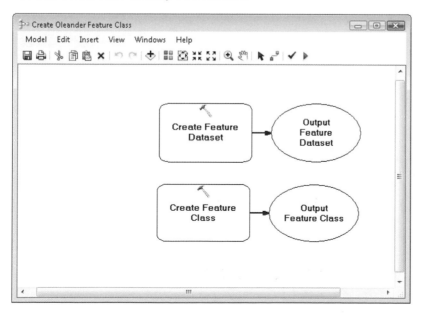

3 On the ModelBuilder menu bar, click Insert > Create Variable to add a new stand-alone variable. Set the data type to Workspace and click OK. This will be the geodatabase where the new feature dataset will be created.

4 Double-click the new variable and set its value by navigating to C:\ESRIPress\ GTKModelbuilder\MyAnswers and selecting the Results geodatabase.

5 Create another stand-alone variable with a data type of String and make it a model parameter. Rename it **New FDS**.

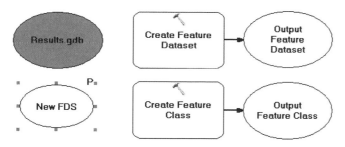

6 Double-click the variable New FDS and type **New Feature Dataset** as the default value. Click OK.

7 Next, create the last stand-alone variable with a data type of String, make it a model parameter, and rename it **New FC**.

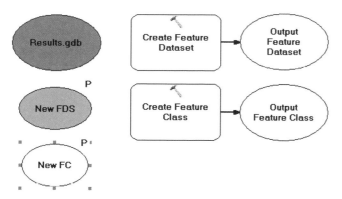

8 Double-click the New FC variable and set the default value to New Feature Class. Click OK, and then save the model.

You may wonder why the first variable is a specific feature type of workspace, but the other two are a generic string type. This is because the geodatabase that will contain the new feature dataset must already exist, so the user prompt must be for a geodatabase. The other two variables will be used to create new elements, so the user prompt should ask for a name and not an existing element. If you wanted the user to create the new feature class in an existing feature dataset, the variable type would have to be set to that.

All the necessary variables have been created, so the next step is to start filling in the tool parameters.

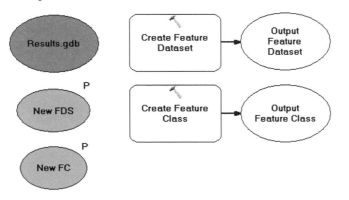

9   Double-click the Create Feature Dataset tool. Set the feature dataset location to Results.gdb. **Tip:** the blue recycle icon next to the dataset indicates that this value is coming from variables already contained in the model.

10   Type the feature dataset name **%New FDS%** as the stand-alone variable. This substitutes the variable into the value box.

Create Feature Dataset

Feature Dataset Location
Results.gdb

Feature Dataset Name
%New FDS%

Coordinate System (optional)

OK   Cancel   Apply   Show Help >>

11   Finally, use the coordinate system Browse button to set this value. Click Import and browse to the City of Oleander geodatabase, and then look in the Accident Information feature dataset for the AccidentTemplate feature class. Select it and click Add. Then click OK.

12 All the parameters for the tool are set. Click OK.

13 Rename the output variable for the Create Feature Dataset tool **Created FDS**. Save the model.

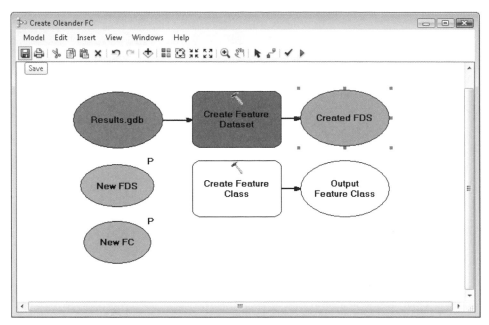

Several things happened when you set the parameters for the Create Feature Dataset tool. A connector line was added to link the geodatabase as a parameter for the tool, and the tool was moved to the Ready to Run state. There is, however, still a tool in the Not Ready to Run state that must be configured.

14 Double-click the Create Feature Class tool. Then set the feature class location to Created FDS.

15 Next, type the feature class name **%New FC%** as the stand-alone variable.

16 Set the geometry type to Point, and set the template feature class to Accident Schema. **Tip:** the yellow symbol next to Accident Schema indicates that it is coming from the table of contents and not from within the model. Click OK.

17 Right-click the Create Feature Class tool, and then click Properties. Go to the Preconditions tab and select the Created FDS check box. Click OK.

18 Click the center of the connector line between the Created FDS variable and the Create Feature Class tool, and then drag the mouse down a bit. Note the dashed line that connects the variable and the tool. This indicates a precondition—that is, the variable must exist before the tool can run. Save the model and close the model window.

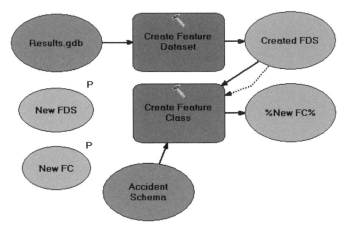

19 In the Catalog window, double-click the Create Oleander FC model to run it. Note that the names you entered for default values serve as a prompt for what the user should type into each input box. Enter your name for the feature dataset and type **PD_3344** as the name for the feature class. Click OK.

The precondition ensures that the feature dataset is created before an attempt is made to create the new feature class. A quick look in the Catalog window shows that the new elements were created as expected.

Even though this model seems linear, it was a safe move to use a precondition when working with the stand-alone variables. Later exercises use preconditions in nonlinear models.

What you've learned so far

◆ How to use an import function to set the spatial reference of a layer
◆ How to use in-line variable substitution
◆ How to set a precondition and control the order in which processes run

Using if-elif-else statements

Flow of control also includes having the model make a decision, or evaluate a situation and decide which step to do next. ModelBuilder has a variety of tools, known as Model Only tools, to handle this type of functionality. The first Model Only tool is called Calculate Value.

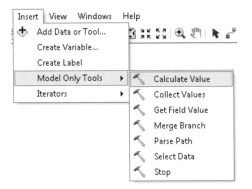

Model Only tools are designed to run exclusively inside models.

The Calculate Value tool evaluates whatever input you give it and produces a single output. It may be a simple equation or a Python script. An included script may call any Python function to evaluate the input, but it is important to remember that only one output is produced. Variable substitution can be used within the expression that this tool evaluates. In the accompanying example, the Calculate Value tool accepts two input values using variable substitution and multiplies them. The output is the multiplied value.

This calculation uses mathematical operators.

Optionally, a block of Python code can be added to perform a more complex function. Note that the code block section of the tool must define a function to accept a value from the expression line. Multiple inputs can be accepted, but remember that only one output is produced. In this example, an input feature class is examined and its feature type is returned.

Calculations can also include Python code.

The Calculate Value tool can be used in conjunction with another Model Only tool, the Stop tool, to create an if-elif-else control. The Stop tool evaluates an input Boolean value and can cue the model to either stop running or continue with another process, depending on the value. Creating an expression in the Calculate Value tool that returns a Boolean value, and then sending that value to the Stop command, is a way to incorporate an if-elif-else control in a model. In the accompanying example, an input value is checked to see whether it is greater than 0. If so, the model continues running; otherwise, the Stop command stops the model.

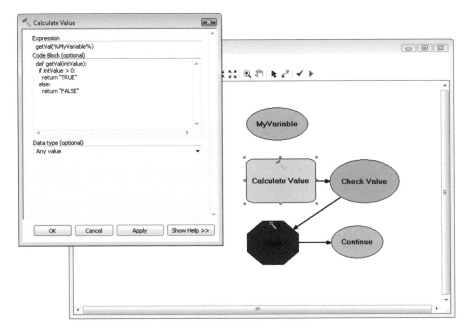

This Python code includes an if-elif-else control.

The Calculate Value tool can be used to determine whether a feature class already exists, whether the correct number of features is selected, or to confirm any condition that can be checked in a Python script. The results are then used to cue the model to either continue running or stop running.

In exercise 2b using the accident location data, you started the model assuming that the user had already selected some features. In the public works notification model you created in exercise 2c, you used a feature set variable to allow the user to select features, and then act on them. What would happen to these models if no features were selected? In some cases, an empty tool output would cause the model to crash, and in other cases, it might act on all the features in a layer. Either way, the results would not be right.

In the following exercise, the staff in public works is using the Public Works Notify tool you created for them, but they are having problems when the lines they draw result in an empty selection set. The model generates error messages, and an empty output table is created. Since the staff is not trained in using the Catalog window for data management, they cannot delete the empty table so they must run the model again using a less desirable name. Rather than deal with this scenario, they want the model to give them some idea of the problem without producing empty tables. The public works director has asked whether there is a way to modify the model and avoid this error. In this exercise, you will take the model and add tools to see how many features are selected. If the number is 0, the model will stop, and no output table will be generated. Any other result will cue the model to continue.

Before you begin the exercise, examine the steps needed to complete the task:

- Test the model with no features selected and inspect the error message.
- Start editing the model.
- Add the Get Count, Calculate Value, and Stop tools to the model.
- Create an if-elif-else process in the model.
- Set a precondition for the Select Layer By Location tool.
- Run the model with no features selected and note how errors are handled.
- Run the model again with features selected and note the results.

Exercise 3b

1 Start ArcMap and open EX03B.mxd. Copy the toolbox called Chapter 3b from the SampleModels folder to your MyAnswers folder.

2 To see what happens when the model runs with no features selected, double-click the Public Works Notify model. Set the output file to C:\ESRIPress\GTKModelbuilder\ MyAnswers\Results.gdb\Empty_Output_Table, and then click OK without selecting any features.

3 The resulting message shows that an empty output file was generated. This is what is causing the concern in public works, and what they are trying to avoid. If you did not have a Results window open on your screen, you can see the messages by clicking Geoprocessing > Results on the ArcMap main menu bar.

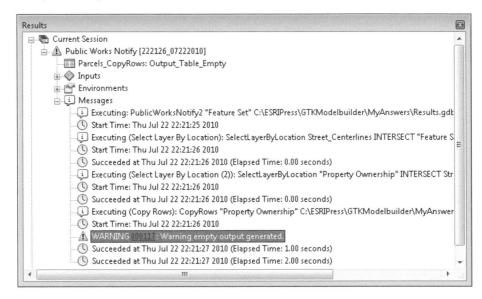

The solution is to create an if-elif-else control with the Check Value and Stop tools, based on the results of the Get Count tool.

4 Start editing the Public Works Notify script in the model window. Use the Search window to locate and add the Get Count tool to the model. On the ModelBuilder menu bar, click Insert > Model Only Tools > Calculate Value, and then repeat the process to add the Stop tool. Arrange these tools along the bottom of the model.

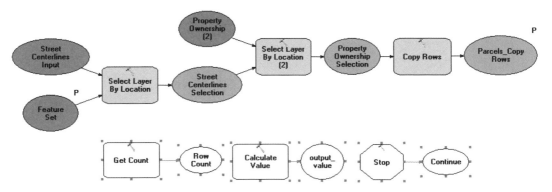

5 Use the Connect tool to connect the output of the Calculate Value tool to the Stop tool with the value set as input values.

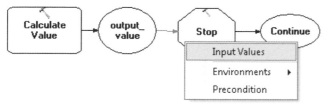

3

4

6 Next, use the Connect tool to make the Street Centerlines Selection variable the input rows for the Get Count tool.

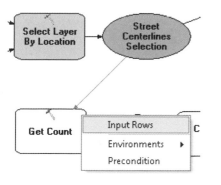

In order for the Calculate Value tool to function as an if-elif-else control, you need to add Python code to the tool's code block. The tool then runs the code and sends the result to the Stop command. The steps in the code start a Python function to accept the count of selected features, and then use an if statement to see whether it is OK (then) to continue. Finally, it uses an else statement to produce a value if the first value does not evaluate to true.

7 Double-click the Calculate Value tool and type the code shown in the following graphic into the code block pane, with each statement on a new line. Press ENTER at the end of each line. Note that the code honors all the Python rules, including indentations and the addition of descriptions. The code, however, does not indent with tabs, so you must type spaces to indent. The if statement is indented two spaces, and the return statement is indented four spaces.

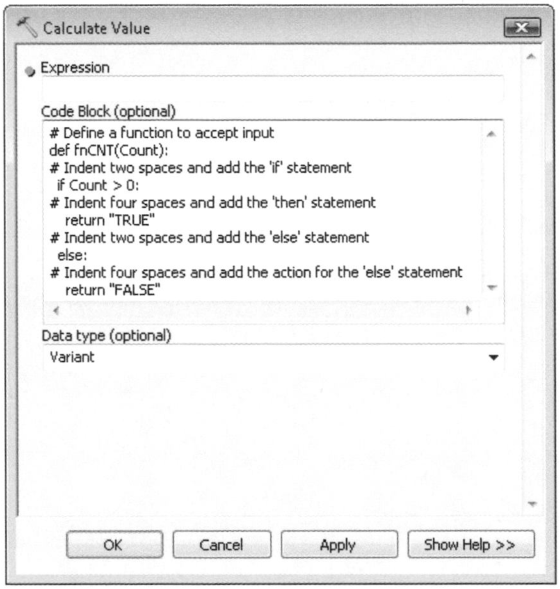

8 Next, add the expression as shown. Note that this uses variable substitution to bring in the result of the Get Count tool and supply it to the Python function in the code block. Click OK to close the Calculate Value dialog box.

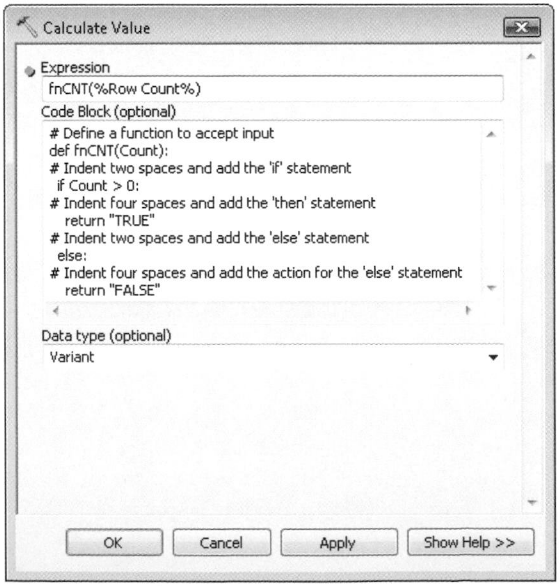

You will need to initialize the output value of the Calculate Value tool so that it can be set in the Stop tool.

9 Right-click the Calculate Value tool, and then click Run. When it has finished running, close the Results window if necessary.

Note that the Calculate tool has moved to the Has Been Run state and the output value has, too. This sets a default value for the Stop command.

10 Double-click the Stop tool and set the "Stop when inputs are (optional)" line to false. This is the value that the Stop tool evaluates.

Finally, there are two preconditions to set to make sure the tools run in the proper order. The first is a precondition set on the Calculate Value tool so that it runs after the Get Count tool, and the second is a precondition set on the Select Layer By Location (2) tool to make sure it doesn't run until the Stop command determines features are selected.

11 Open the properties of the Select Layer By Location (2) tool and click the Preconditions tab. Select the Continue check box and click OK. Then open the properties of the Calculate Value tool and set the precondition to Row Count. Save the model and close the model window.

As the model runs, the feature set variable prompts the user to draw a line on the map. That line is used by the model to select underlying street centerline features with the Select Layer By Location tool. The model feeds the output of that process into the Get

Count tool to see whether any features were, in fact, selected. The Calculate Value tool determines whether the number selected is greater than 0 and passes a corresponding true or false value to the Stop tool. Then the Stop tool cues the model to stop if the value is false or continue if it is true. Note the dashed connector lines showing the preconditions you set.

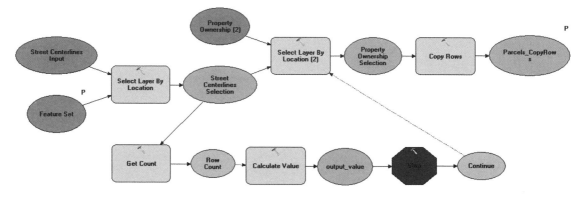

12 To test the model, double-click it and enter **C:\ESRIPress\GTKModelbuilder\MyAnswers\ Results.gdb\Empty_Output_Table_2** as the output table. Then click OK without selecting any features.

13 You'll see that the model didn't run. The messages in the Results window state that there were no features selected and that the second selection process didn't run because the precondition was false. If you examine the Catalog window, you'll find that no empty output table was created.

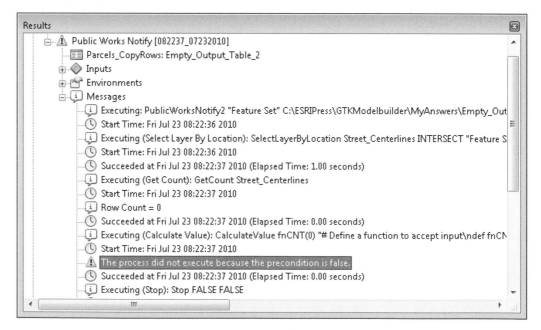

14 Run the Public Works Notify tool again, and this time, use the Add Features tool to draw a line over Peterstow Drive. Set the output table to **C:\ESRIPress\ GTKModelbuilder\MyAnswers\Results.gdb\Peterstow** and click OK.

The model ran correctly and the output table Peterstow was created. The addition of decision-making controls within the model fixes the problem that occurs when no features are selected.

Error checking is a big concern in designing models. If the model is testing flow or predicting optimal routes, the user needs to determine what data anomalies might cause a process to fail and stop the model unexpectedly. With error-checking procedures built in, a model can be run with confidence to simulate a number of scenarios. The next two scenarios in this chapter also implement error checking to test for data irregularities.

What you've learned so far

- How to add error checking to a model
- How to add an if-elif-else control to a model
- How to inspect errors generated in a model

Branching and merging

Flow of control also includes having the model evaluate a situation and decide on one of several courses of action. This is called branching. The process involving the Stop tool only had one branch to follow, or else it stopped. Using a Python script, you can indicate any number of branches to follow after an evaluation statement. The format follows a pattern like this:

```
IF some condition exists

    Do some process (branch 1)

ELSE do a different process or quit (branch 2)
```

The code in a script would look something like the accompanying graphic.

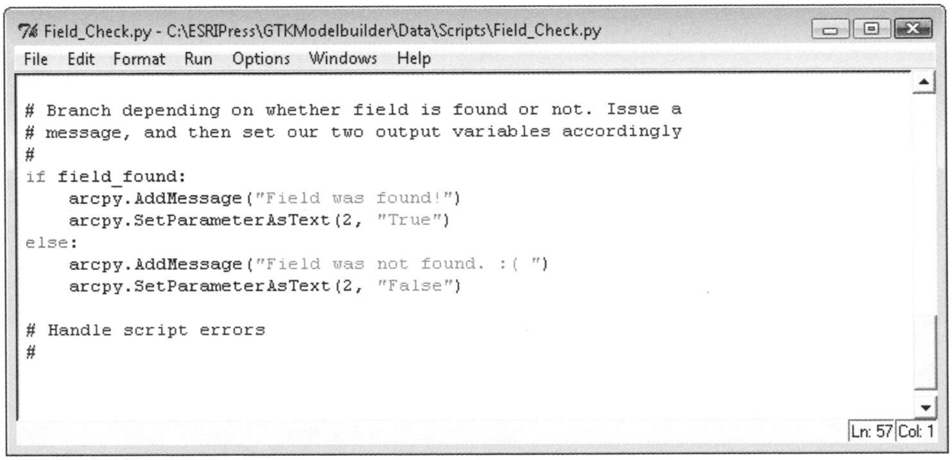

```
7 Field_Check.py - C:\ESRIPress\GTKModelbuilder\Data\Scripts\Field_Check.py
File  Edit  Format  Run  Options  Windows  Help

# Branch depending on whether field is found or not. Issue a
# message, and then set our two output variables accordingly
#
if field_found:
    arcpy.AddMessage("Field was found!")
    arcpy.SetParameterAsText(2, "True")
else:
    arcpy.AddMessage("Field was not found. :( ")
    arcpy.SetParameterAsText(2, "False")

# Handle script errors
#
                                                        Ln: 57 Col: 1
```

Python script containing a branch process

The script is used to see whether a field already exists in the table. If the field is found, a message is sent to the user and the output variable is set to true. If it is not found, the else statements are executed so that a different message is given and the output variable is set to false. When the script is brought inside ModelBuilder, the outcomes with their connector lines look like branches of a tree or a stream.

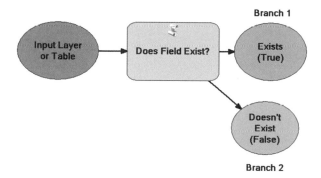

The branch is shown visually in ModelBuilder.

This type of script has only two branches, but with the use of the Python if-elif-else command, it could have many branches. The format is as follows:

```
IF some condition exists

      Do some process (branch 1)

ELIF check a different condition

      Do a different process (branch 2)

ELIF check a different condition

      Do a different process (branch 3)

ELIF check a different condition

      Do a different process (branch 4)
```

You would repeat the elif command for each branch. In a script, it looks like the accompanying graphic.

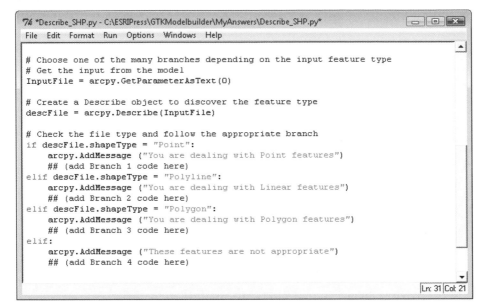

```
# Choose one of the many branches depending on the input feature type
# Get the input from the model
InputFile = arcpy.GetParameterAsText(0)

# Create a Describe object to discover the feature type
descFile = arcpy.Describe(InputFile)

# Check the file type and follow the appropriate branch
if descFile.shapeType = "Point":
    arcpy.AddMessage ("You are dealing with Point features")
    ## (add Branch 1 code here)
elif descFile.shapeType = "Polyline":
    arcpy.AddMessage ("You are dealing with Linear features")
    ## (add Branch 2 code here)
elif descFile.shapeType = "Polygon":
    arcpy.AddMessage ("You are dealing with Polygon features")
    ## (add Branch 3 code here)
elif:
    arcpy.AddMessage ("These features are not appropriate")
    ## (add Branch 4 code here)
```

This Python script contains multiple branches.

And in a model, it looks like the accompanying graphic.

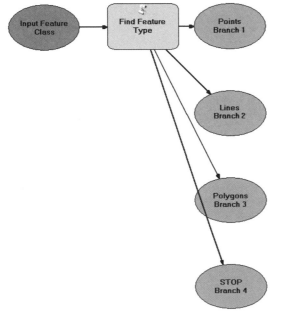

Python script visualized in ModelBuilder

Note that the script appears in the ModelBuilder diagram as a tool, but it has a different icon showing that its origin is a script and not an ArcGIS geoprocessing tool.

Having the model make a decision based on this branching technique also involves setting a precondition. The branch variable can be set as a precondition for a tool, so that the tool runs only when the variable is true.

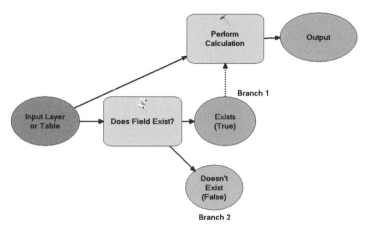

Model process triggered by true value

Branches may also be used to steer a model to use different tools. For example, if the user input for a model is point features, that branch might call a point tool. However, if linear features are selected, the second branch might call a linear tool. By setting preconditions on the tools, control passes to different model processes based on the input.

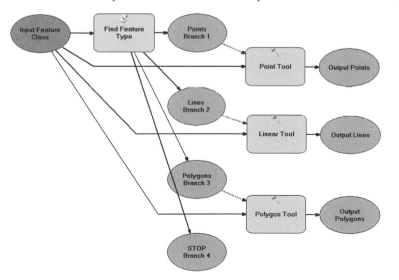

The model detects different feature types and passes control to the appropriate process.

Branch operations can be used to determine whether the right number or type of features is selected, whether the proper calculations have been made in a process, whether the results fit within the project parameters, or basically anything that can be calculated in a script. This is a very powerful component of ModelBuilder, giving it the capabilities of a true programming language.

Decisions that send a program in two different directions may need to be brought back together to finish the project. In the preceding example, the next step may be to buffer the output. Since the Buffer tool works on points, lines, or polygons, the results of any branch can be fed into the Buffer tool. To bring the processes back together after the branch, the Merge Branch tool is used. This tool exists for exclusive use inside ModelBuilder to return the model to a linear path. The tool takes multiple input variables and outputs the first variable that reaches the Has Been Run state.

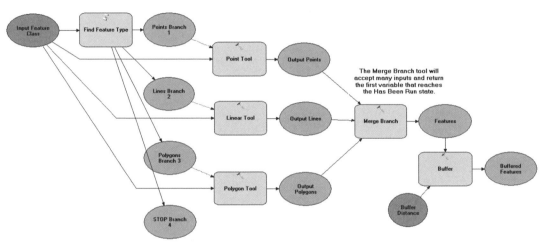

Use of the Merge Branch tool

Once the model is back on a linear path, more tools or scripts can be added. Another decision-making branch can also be created and added to the model.

Understanding the ArcPy Python module

The use of scripts for decision making requires that the script be able to extract information from the model components. A full set of geoprocessing operators can be added to a Python script to give the programmer the ability to discover a wealth of information for use in various processes and decision-making routines. In the following script example, the first line of code loads this set of ArcGIS operators, or modules, called ArcPy. Other modules, such as the OS and Sys modules, which contain native Python operators, are also loaded.

```
# Import system modules
# Add the ArcGIS tools with the ArcPy module
import arcpy, sys, os
```

The ArcPy module adds geoprocessing tools to Python.

Once the ArcPy module is loaded, the ArcGIS geoprocessing tools become available as methods inside the Python script. A search of ArcGIS Desktop Help for the keyword "arcpy" leads to a more complete description of what the ArcPy module does and the methods available once it is imported into a Python script.

ArcGIS Desktop Help lists information about the ArcPy module

The ArcPy module, along with any of the other native Python modules, can be loaded and used in a script, and the resulting information can be used in a decision-making process. Each geoprocessing tool shown has an associated Help document that details how the tool works, as well as code samples that demonstrate how the tool can be incorporated into a Python script. For instance, the Help page for the Describe tool, which is used later in this chapter, gives explanatory information and code samples.

Describe

Summary

Returns properties for specified data elements, such as Tables, Feature Classes, Geodatabases, Rasters, Coverage Feature Classes, Layer Files, Relationship Classes, Workspaces, and Datasets, as well as geoprocessing objects such as FeatureLayers and TableViews.

Discussion

The returned value of Describe is an object containing properties, such as data type, fields, indexes, and many others. Its properties are dynamic, meaning that depending on what data type is described, different describe properties will be available for use.

Describe properties are organized into a series of property groups. Any particular dataset will acquire the properties of at least one of these groups. For instance, if describing a geodatabase feature class, you could access properties from the GDB FeatureClass, FeatureClass, Table and Dataset property groups. All data, regardless of the data type, will always acquire the generic Describe Object properties (baseName, catalogPath, children, childrenExpanded, dataType, extension, file, fullPropsRetrieved, metadataRetrieved, name, path).

Describe Object Properties
CAD Drawing Dataset Properties
Coverage FeatureClass Properties
Coverage Properties
Dataset Properties
FeatureClass Properties
GDB FeatureClass Properties
GDB Table Properties
Geometric Network Properties
Layer Properties
Network Analyst Layer Properties
Network Dataset Properties
Prj File Properties
Raster Band Properties
Raster Catalog Properties
Raster Dataset Properties
RelationshipClass Properties
RepresentationClass Properties
Table Properties
TableView Properties
Tin Properties
Topology Properties
Workspace Properties

Syntax

Describe (value)

Parameter	Explanation	Data Type
value	The specified data element or geoprocessing object to describe.	String

Return Value

Data Type	Explanation
Describe	Returns an object with properties detailing the data element or geoprocessing object. Some of the returned object properties may contain literal values or objects. For example, describing a feature class will return information on the feature class, including the fields list object and the extent object. These objects can be be examined further to retrieve each field in the feature class or the XMin,YMin,XMax, or YMax properties of the extent.

Code sample

Evaluate data properties of a geodatabase feature class.

```
import arcpy

# Get the feature class to describe (i.e. "C:/Data/airport.gdb/
#
featureClass = arcpy.GetParameterAsText(0)
desc = arcpy.Describe(featureClass)

# Print selected feature class properties
#
print "shapeType", desc.shapeType
print "hasSpatialIndex", desc.hasSpatialIndex
print "extent XMin", desc.extent.XMin
print "extent YMax", desc.extent.YMax
print "the first field's name", desc.fields[0].name
print "the first field's type", desc.fields[0].type
```

The Help page for the Describe tool lists useful information.

The Help page for the Describe tool is quite detailed on all the functions it can perform, but the most important of these functions have to do with acquiring information about datasets. Lists of fields, extents of coverage, definition queries, and many other types of information can be extracted from feature classes, layers, and tables, which could then be used in scripts or model processes.

The uses of these tools are too numerous to detail here, so a few will be demonstrated for their decision-making capabilities in ModelBuilder.

In chapter 1, you wrote a model to create a new feature dataset and feature class for the police department to use. The project was to create the file structure to store data about a traffic accident. In the following exercise, the officers have had a problem with the creation of the feature dataset. Once the dataset was created, the model would no longer run because it didn't check to see whether the name that was entered already existed. The police chief has asked for some file-checking procedures to prevent this problem.

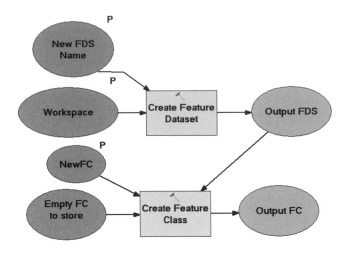

This model for the police department lacks error-checking procedures.

As you can see from the accompanying graphic, the model prompts the user for the workspace in which to create the new elements, the new feature dataset name, and the new feature class name. In the exercise, you'll add a script that checks for the existence of the feature dataset before trying to create it. Then you'll do the same for the feature class. Note that this requires a multibranch process, so a Python script is required rather than using the Calculate Value and Stop tools.

Before you begin the exercise, examine the steps needed to complete the task:

- Review the description of the Exists function in ArcGIS Desktop Help.
- Examine the code in a Python script file.
- Start editing the model.
- Add a Python script to the model.
- Set parameters for the Python script.
- Create a precondition.
- Use the Auto Layout and Full Extent tools.
- Search for geoprocessing tools.
- Merge two branches of the model.
- Validate the model.
- Export the model diagram to a graphic.
- Run and test the model.

Exercise 3c

1 Start ArcMap and open EX03C.mxd. Copy the toolbox called Chapter 3c from the SampleModels folder to your MyAnswers folder. Then review the ArcGIS Desktop Help page for **Exists** to get an idea of how this method is used in a script.

2 In the Chapter 3c toolbox, right-click the Check Exists script and click Edit. An editor window and a Python shell window open. You can close the Python shell window, because you will be running the script in ModelBuilder.

```
# -------------------------------------------------------------
# CheckExists.py
# Created on: Wed Jan 3 2008
# Updated on: Wed Mar 17 2010
#
# -------------------------------------------------------------

# Import system modules
import arcpy, sys, string, os

# Get the input in the model
InputFDS = arcpy.GetParameterAsText(0)
Workspace = "C:/ESRIPress/GTKModelbuilder/MyAnswers/Results.gdb/"
Pathname = Workspace + InputFDS
arcpy.AddMessage(Pathname)


# Statement to check the existence of the feature dataset
if arcpy.Exists(Pathname):
    arcpy.SetParameterAsText(1,Pathname)
    arcpy.AddMessage("Feature Dataset Already Exists!")
else:
    arcpy.SetParameterAsText(1,"")
    arcpy.SetParameterAsText(2,"True")
    arcpy.AddMessage("Feature Dataset Does Not Exist!")
```

CheckExists.py - C:\ESRIPress\GTKModelbuilder\Data\Scripts\CheckExists.py

File Edit Format Run Options Windows Help

Ln: 1 Col: 0

Follow the accompanying code for this script. You can see at the top that it imports the ArcPy module. Then it creates an input variable to hold the user's suggested feature dataset name. This is concatenated with the path for the feature dataset and stored in a new variable called Pathname. The last part runs the arcpy.Exists method, which tests for the existence of the specified data object and checks the input feature dataset name, which is stored in the Pathname variable. If the item exists, the valid path is passed on to an output variable. Your model simply takes this value and proceeds to the Create Feature Class tool. If the item does not exist, a Boolean value of false is passed on to an output variable. This is used as a precondition to tell the model that a new feature dataset must be created.

3 Close the script editor and start editing the Create Oleander FC model. Zoom out a bit to make some room in the model window, and then drag the Check Exists script into the model.

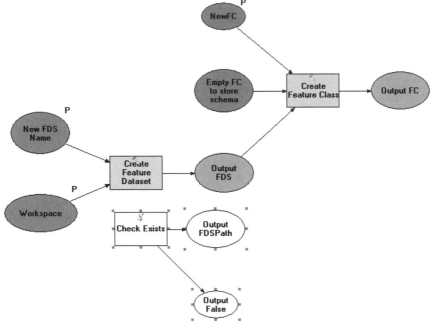

4 Double-click the Check Exists script and set Input FDS Name to New FDS Name. Click OK and note that the script moves to the Ready to Run State. Click the Auto Layout button to organize the model elements.

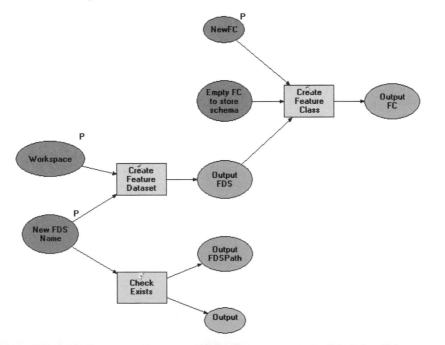

5 Select the connector line between the Output FDS variable and the Create Feature Class tool. Delete it by right-clicking the connector line and then clicking Delete on the context menu or by pressing DELETE on your keyboard and answering Yes to the prompt. Note that the Create Feature Class process moves to the Not Ready to Run state.

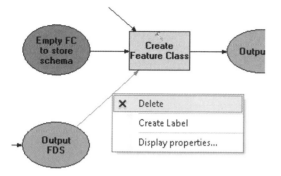

6 Open the properties of the Create Feature Dataset tool and go to the Preconditions tab. Select the Output False check box. Click OK. Now this tool won't run unless the output of the Check Exists script is false, meaning that the user's suggested feature dataset does not exist.

7 The diagram is getting a little messy. Click the Auto Layout button, and then click the Full Extent button to clean up the display.

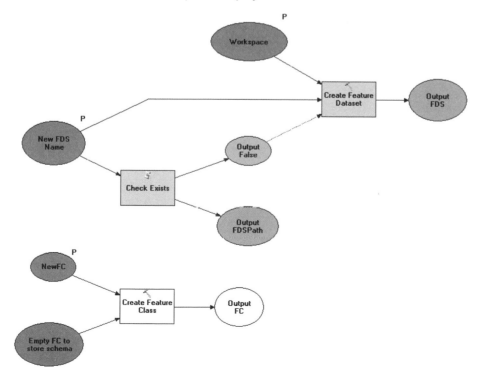

8 On the ModelBuilder menu bar, click Insert > Model Only Tools > Merge Branch and drag the new tool to the lower right corner of the model.

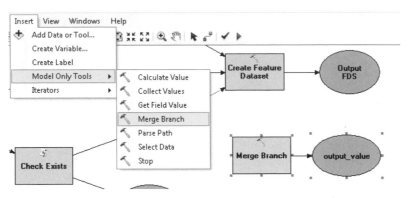

9 Double-click the Merge Branch tool to open its dialog box. In the In Values text box, add the variables Output FDS and Output FDSPath. Make sure the variables are in this order. If not, reverse them. Click OK.

The Merge Branch tool looks at both inputs, in the order they are listed, and determines whether either input contains a valid value. The first input on the list with a valid value is output to the next process.

10 Rename the output variable from the Merge Branch tool **Merge FDS** and use the Connect tool to connect it to the Create Feature Class tool. Set it as the feature class location.

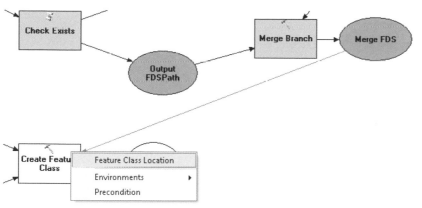

11 Click the Auto Layout button and then the Full Extent button if necessary to see the entire model diagram.

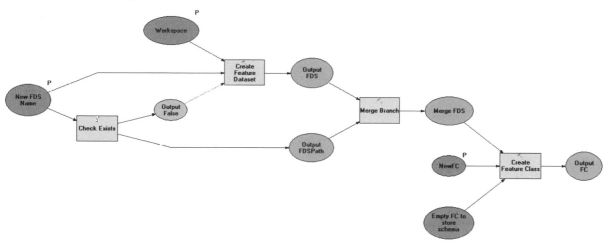

12 If you like, try making an output graphic of the model and view the model report. Save the model and close the model window.

It would also be desirable to have the model prompt the user for input in this order: workspace, feature dataset, feature class. You can control this order in the model's properties window.

13 In the Catalog window, right-click the Create Oleander FC model, and then click Properties. Click the Parameters tab to see all the inputs for this model. Place them in the correct order.

Create Oleander FC Properties

| General | Parameters | Environments | Help | Iteration |

Parameters used by this model:

Name	Data Type	Type	Filter
Workspace	Workspace	Required	None
New FDS Name	Any value	Required	None
NewFC	Any value	Required	None

OK   Cancel   Apply

14 Test the functionality of your model by running it with new values. For the first test, type a new name for the feature class and the name of an existing feature dataset. Note the results. In the second test, provide a new name for both the feature class and the feature dataset.

This prevents errors if a feature dataset already exists and creates a new feature dataset if the name given does not exist.

What you've learned so far

- ◆ How to identify the components of a Python script
- ◆ How to set script parameters
- ◆ How to adjust the display of the model components within the editing dialog box
- ◆ How to merge several branches of a model
- ◆ How to validate the components and their parameters in a model

Creating a basic Python script

The exercises so far demonstrate the usefulness of scripts in models. But where does the code come from and how is it integrated into a script?

Many code samples, along with tutorials, can be found in ArcGIS Desktop Help. Others are available on the Geoprocessing Resource Center page where ArcGIS users share code for various processes (http://resources.arcgis.com/geoprocessing/). But the most common way to work with scripts is probably to write scripts from scratch or use snippets of code from other sources. The scripts you use in your models can be in Python or other programming languages, but the code samples in ArcGIS Desktop Help are all in Python, so that's what is used in this book.

Python scripts are created in the code editor loaded along with ArcGIS. The code editor is called IDLE and can be run inside ArcMap with the Python window. Other Python code editing software, such as PythonWin, which comes on the ArcGIS installation disk, can be used outside ArcMap but is dependent on the operating system.

The basic Python script to be used in a model does the following:

- Loads the ArcPy module
- Establishes variables to receive input (based on the geoprocessing task)
- Performs an ArcGIS geoprocessing task
- Evaluates an if-elif-else statement (optional)
- Presents output variables or messages

Scripts should also contain comment lines to document each step. In Python, these lines are preceded by the pound sign (#). They are useful to the script creator as a reminder of the process steps and are invaluable to other programmers as an explanation of what each step is doing.

In the accompanying example, code is written to look at a user-specified table and check for the existence of a user-specified field. The script is called Field_Check.py and is included under Scripts in the Data folder on the accompanying DVD.

```
7⁄ Field_Check.py - C:\ESRIPress\GTKModelbuilder\Data\Scripts\Field_Check.py
File  Edit  Format  Run  Options  Windows  Help
#**********************************************************************
# Description:
# Tests if a field exists and outputs two booleans:
#   Exists - true if the field exists, false if it doesn't exist
#   Not_Exists - true if the field doesn't exist, false if it does exist
#                (the logical NOT of the first output).
#
# Arguments:
#   0 - Input Table name (required)
#   1 - Input Field name (required)
#   2 - Output Boolean Variable for Exists (see above)
#   3 - Output Boolean Variable for Not_Exists (see above)
#
# Created by: ESRI
# Fall 2008
# Updated Spring 2010
# ArcGIS 10 using arcpy module
#**********************************************************************
```

Python scripts should contain documentation on their inputs, outputs, and processes.

The first section contains documentation about the purpose of the model, as well as the expected outputs. It also contains a list of arguments for both the input and output variables. These are helpful when the script is brought into ModelBuilder. The last few lines list the author, a date of creation, and a reference to the software version for which the model is written. This information can be useful if the model is updated.

```
# Import system modules
import arcpy, sys, string, os
```

These lines load the ArcPy module and some native Python modules.

The next section loads the ArcPy module. This module contains all the geoprocessing and file maintenance tools written by Esri. This is sensitive to the version of ArcGIS that you are running, because as new tools are added to ArcGIS they are also included in the ArcPy module. In ArcGIS 10 software, each of the geoprocessing tools is prefaced by "arcpy." For example, the Exists tool in the previous exercise is noted in the Python script as "arcpy.Exists." The Sys, String, and OS modules that are imported contain basic Python tools. These can be found in a good Python reference book or at http://www.python.org.

```
# Get input arguments - table name, field name
#
in_Table = arcpy.GetParameterAsText(0)
in_Field = arcpy.GetParameterAsText(1)
```

These lines accept input for the script.

After the modules are loaded, the user input fields are created. The input fields, described in the initial section, use the GetParameterAsText tool. This is a special input statement for use with ArcGIS scripts and must be preceded by the "arcpy" prefix. Another more generic system argument statement (sys.argv[ ]) may appear in several of the ArcGIS Desktop Help examples. This is another way to input data using a standard Python module rather than one specifically created for geodata. Each input is given an index number with the first being 0, the second being 1, the third being 2, and so on. Each Python statement starts with 0 and goes up to however many inputs there are. **Tip:** if the sys.argv[ ] statement is used, the first index number is 1 and goes forward from there.

```
# First check that the table exists
#
if not arcpy.Exists(in_Table):
    raise Exception, "Input table does not exist"
```

The if control is used for error checking.

In the accompanying example, a quick if statement is added to make sure the table exists. The geoprocessing task Exists is used here, which returns true for existing tables. Note that by using "not" as a condition of the if statement, the process is reduced to two lines. It is basically saying, if the table does not exist, type a warning and quit. There is no then statement, because the process continues automatically to the next step if the table exists.

```
# Use the ListFields function to return a list of fields that matches
#  the name of in_Field. This is a wildcard match. Since in_Field is an
#  exact string (no wildcards like "*"), only one field should be returned,
#  exactly matching the input field name.
#
fields = arcpy.ListFields(in_Table, in_Field)

# If ListFields returned anything, the Next operator will fetch the
#  field. We can use this as a Boolean condition.
#
field_found = fields.Next()
```

Information is extracted from the input datasets.

In the next step, a task extracts the list of field names from the table and looks for the specific target field. This task uses the geoprocessing tool List Fields. When the field's Next() statement is issued, the field_found variable either is blank if the field is not found or contains the field name if the field is found.

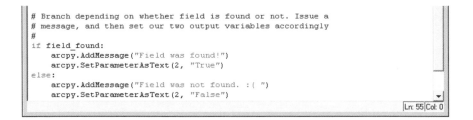

```
# Branch depending on whether field is found or not. Issue a
# message, and then set our two output variables accordingly
#
if field_found:
    arcpy.AddMessage("Field was found!")
    arcpy.SetParameterAsText(2, "True")
else:
    arcpy.AddMessage("Field was not found. :( ")
    arcpy.SetParameterAsText(2, "False")
```
```
                                                            Ln: 55 Col: 0
```

This if-elif-else control manages branching.

The List Fields tool, however, does not indicate where the process should go next. For this, an if-elif-else statement is used. If the field exists, the if statement evaluates to true and the first set of code is processed. This prints a message stating that the field is found. The next two statements create the output for the script. The SetParameterAsText tools create the output that appears in the model. Note that the tool arguments use index numbers that continue counting upward from the input variables, and then supply a value of true or false. For situations when the field does not exist, the second set of code is processed. Note that the outputs use the same index numbers, but the values are reversed.

This script is ready to run in Python, but it is not yet ready to run in a model. For that, the script needs to be imported into a toolbox and the input and output variables defined. The first step is to add the script to a toolbox.

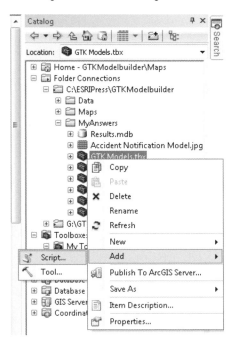

An ArcPy script is added to a toolbox.

The Add Script dialog box asks for a script name, a label, and a description. Other options include a style sheet, which lets users customize the look of their input windows, and the option to store relative path names rather than absolute paths. Storing relative paths makes the script easier to share, because others receiving the script do not have to use identical path structures.

A script contains core information when it is added to a toolbox.

The next step is to supply the location of the source code file. The .py extension refers to a Python script. The option to run Python scripts "in process" means that the script runs in the ModelBuilder process that calls the script rather than the script invoking a new process outside the model. Running in process saves time and can help eliminate difficulties in transferring messages from one process to another.

The script file is identified and the process state is set.

The final step is to set up the input and output variables that the script will use. The script's code contains a list of the variables and their index numbers. The variables are then added as script parameters in the order of their index numbers. A display name is given, and the data type is set. The first two variables (index 0 and 1) are listed as input variables with their direction parameter set to Input and their type set to Required. If the data types were set to Table and Field, respectively, the user would be given navigation dialog boxes when the model is run that would accept only these types of files. Since the output of this script may cause these things to be created, the variables are set to "Any value" so that the user does not face any restrictions.

The variable type is selected.

The last two variables (index 2 and 3) are listed as Boolean (true/false) variables with their direction parameter set to Output. Their type is automatically set to Derived. Output variables may also contain values of any of the data types listed, and these values are passed on to the model.

Display Name	Data Type
Input Table Name	Any value
Input Field name	Any value
Output True	Boolean
Ⓒ Output False	Boolean

Click any parameter above to see its properties below.

Parameter Properties

Property	Value
Type	Derived
Direction	Output
MultiValue	No
Default	
Environment	
Filter	None

Variables are noted as input or output.

This script is now ready to be dragged into a model. The script provides the two output variables in the ModelBuilder diagram. Note also that the element's icon now represents a script rather than a tool.

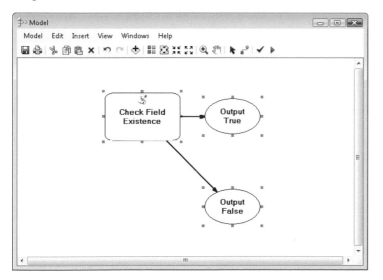

Visual depiction of a script in the model window

The script requires two inputs, a table name and a field name, before it can be moved into the Ready to Run state. **Tip:** the green dots indicate that the fields are required.

The model acts as the conduit to feed the parameters into the script.

The process described here shows how to create a simple script and move it into ModelBuilder. Scripts can also be used for many purposes outside of ModelBuilder. For more information, check the ArcGIS Desktop Help topic "Writing Python scripts," search the Esri Virtual Campus for instructor-led and online Python courses, or review other Python resources.

Exercise 3c built an error-checking mechanism to check for the existence of the user-suggested feature dataset, creating a new one when necessary. In the following exercise, the same problem is occurring with the feature classes. If the user suggests a name of a feature class, and that name already exists, the model crashes. In the exercise, you'll need to add a script that checks for errors, except this time the code isn't already written. Instead, you'll write a new Python script from scratch to accomplish this task. Once the script is created, you'll incorporate it into a model.

Before you begin the exercise, examine the steps needed to complete the task:

- Create a new Python script.
- Write lines of code into the script editor. The new script needs to include the following:
    - Documentation to show the purpose of the script, record a list of variables, and provide source information
    - A step to load the ArcPy module
    - Code to establish input variables and accept user input (there will be a variable to accept the suggested feature class name)
    - An if-elif-else statement to evaluate the geoprocessing tool called Exists
    - Output variables for true and false based on the results of the if-elif-else statement
- Define input variables.
- Set up an if-elif-else routine.
- Save the Python script.
- Add the script to your toolbox.
- Define the tool variable for the script.
- Add the script to a model.

Exercise 3d

1 On the Start menu, click All Programs > ArcGIS > Python 2.6 > IDLE and run it. This opens the Python shell window, but currently there is no code to run.

2 Click File > New Window to open a coding window.

3 Begin the code by typing a row of pound signs (#) and press ENTER. Then type a single pound sign and space and type the name of the file, **FC_Check.py**.

Continue by typing a description of what this script will accomplish. Be sure to start each line with a pound sign.

Next, add the word **Arguments:** with a colon. On the next few lines, add the index numbers and variable names to be used, with one space on either side of the hyphen:

```
0 - Input Feature Class Name (required)

1 - Output Boolean variable for Exists

2 - Output Boolean variable for Does Not Exist
```

Complete the initial documentation area by typing your name, the date this script was created, and the ArcGIS level required to run it. Finish with a row of pound signs.

Exercise **3d** Creating a basic
Python script

```
7% *FC_Check.py - C:\ESRIPress\GTKModelbuilder\Data\Scripts\FC_Check.py*
File  Edit  Format  Run  Options  Windows  Help
###################################################################################
# FC_Check.py
# This script will test for the existence of a user specified Feature Class.
# The ouputs will be either True or False.
#
# Arguments:
# 0 - Input Feature Class Name (required)
# 1 - Output Boolean variable for Exists
# 2 - Output Boolean variable for Does Not Exist
#
# Your name
# Month Year
# ArcGIS 10
#
###################################################################################
#
```

4 The next step is to write code to load the ArcPy module, along with the Sys and OS Python modules if they are to be used. Type the following:

```
# Load the Arcpy module

# This makes all the ArcGIS geoprocessing tools available in Python

import sys, os, arcpy
```

```
# Load the arcpy module
# This makes all the ArcGIS geoprocessing tools available in Python
import sys, os, arcpy
```

Python is case sensitive, so check your code carefully. Check the online Python documentation at www.python.org for lists of the functions associated with the Sys and OS modules.

5 Now add the code to set up the input variables:

```
# Get the input from the model

InputFC = arcpy.GetParameterAsText(0)
```

```
# Get the input from the model
InputFC = arcpy.GetParameterAsText(0)
```

6 Next comes the if-elif-else statement. Start by typing:

```
# Statement to check the existence of the feature class

if arcpy.Exists(InputFC):
```

Press ENTER after typing the colon. The next line is automatically indented. The indentations are how Python determines what is to be included in the if statement. On the indented lines, type:

```
arcpy.SetParameterAsText(1,"True")

arcpy.SetParameterAsText(2,"False")

raise Exception, "Feature Class Already Exists!"
```

Press BACKSPACE to remove the indentation. This marks the end of the if statement.

Next, type:

```
else:
```

Press ENTER after the colon and the next line is again indented. Now type:

```
arcpy.SetParameterAsText(1,"False")

arcpy.SetParameterAsText(2,"True")

arcpy.AddMessage("Feature Class Does Not Exist!")
```

Be careful to mind the indentations and capitalization because Python is very sensitive to these.

```
# Statement to check the existence of the feature class
if arcpy.Exists(InputFC):
    arcpy.SetParameterAsText(1,"True")
    arcpy.SetParameterAsText(2,"False")
    raise Exception, "Feature Class Already Exists!"
else:
    arcpy.SetParameterAsText(1,"False")
    arcpy.SetParameterAsText(2,"True")
    arcpy.AddMessage("Feature Class Does Not Exist!")
```

7 On the Python menu bar, click File > Save As. Then navigate to your MyAnswers folder and save the file as **FC_Check.py.** After the file is saved, close all the Python windows.

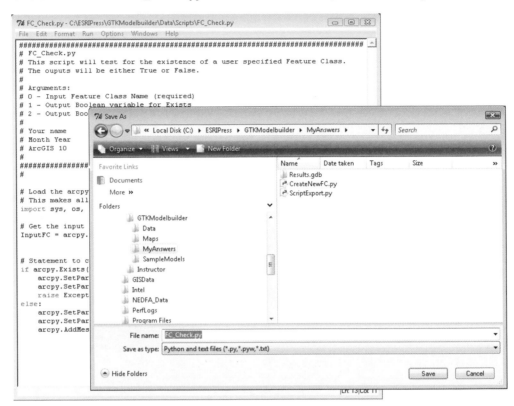

The next step is to add the script to the Chapter 3d toolbox and set up its parameters.

8 Start ArcMap and open EX03D.mxd. Create a new toolbox called **Chapter 3d** in your MyAnswers folder.

9 Right-click the Chapter 3d toolbox, and then click Add > Script.

10 Enter a name of **FCExists**, a label of **Check Feature Class Existence**, and a description of **Check to see if the user-specified feature class exists.** Select the "Store relative path names" check box and click Next.

11 Enter the script file by navigating to the folder where you saved the script and selecting FC_Check.py. Click Open and then click Next.

The last step is to enter the variables. Remember that they must be listed in order of index number, with 0 being first.

12 Click under Display Name and type **Input Feature Class**. Set the data type to any value. Verify that the direction is set to Input and the type is set to Required. These are the defaults.

13 On the next line under Display Name, type **Output True**. Set the data type to Boolean. Change the direction to Output and the type to Derived.

14 Repeat the previous step to create an entry for the **Output False** variable. It is also a derived output variable with a Boolean value. When you have all three variables configured correctly, click Finish. The script now exists in the Chapter 3d toolbox.

Add Script

Display Name	Data Type
Input Feature Class	Any value
Output True	Boolean
€ Output False	Boolean

Click any parameter above to see its properties below.

Parameter Properties

Property	Value
Type	Derived
Direction	Output
MultiValue	No
Default	
Environment	
Filter	None

To add a new parameter, type the name into an empty row in the name column, click in the Data Type column to choose a data type, then edit the Parameter Properties.

< Back   Finish   Cancel

The script is ready to add to the Create Oleander FC model to check for the existence of the user-specified feature class. If the script returns as false, the model continues running and creates the feature class. In the event that the feature class already exists, the script handles the error message and returns control to the model without attempting to create the feature class.

15 Copy the Create Oleander FC model from the Chapter 3d toolbox in the SampleModels folder and paste it into your new toolbox in your MyAnswers folder. Start editing the copy of the model and drag the Check Feature Class script into the model.

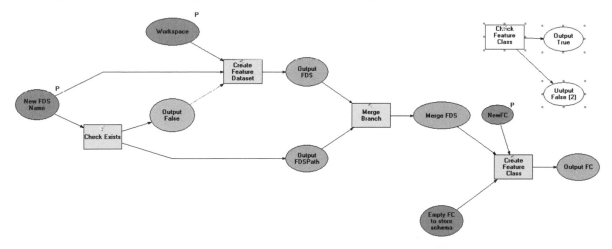

16 Perform the following steps, similar to exercise 3c, to incorporate the script into the model's processes:

   a. Connect the NewFC variable to the Check Feature Class script, making it the input feature class.
   b. Connect the Output False (2) variable to the Create Feature Class tool as a precondition.
   c. Save and close the model.

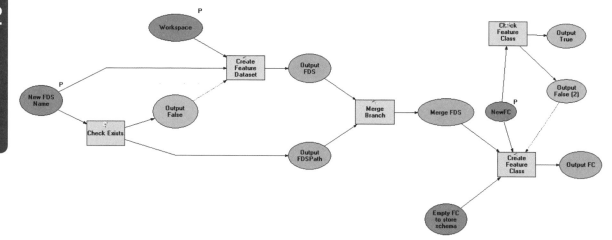

Try testing the model, giving it both an existing feature class and a new feature class. The model now contains error handling for two errors that users might make. This keeps the model from crashing and makes it easier to use.

What you've learned so far

   ◆ How to write a new Python script from scratch
   ◆ How to define input and output variables in a script
   ◆ How to set up an if-elif-else routine in a script
   ◆ How to add a script to a toolbox to create a new script tool
   ◆ How to set up the variables for a script tool in a toolbox

Using specific data types

When variables from scripts are configured in the model environment, they can be set to any one of more than 100 different data types. In the previous exercise, the data type for the input is set to any value. Anything the user types is accepted, and in that particular instance, it was the correct thing to do. In other situations where the user is required to select an existing feature or element, the data type setting inserts a navigation tool appropriate to the selection.

For instance, setting a data type to Spatial Reference requires that the input specifically be an existing spatial reference.

This input is restricted to spatial references.

The input dialog box for this script when run in a model is the standard ArcGIS input dialog box for spatial reference and accepts only a valid spatial reference as an entry.

When the model is run, ArcMap supplies the correct dialog box for selecting a spatial reference.

In addition, some data types can be made conditional on other data types. For instance, setting a variable as a data type of Table will let users select only a table. By adding a second variable with a data type of Field and making it conditional on the first variable, the user is presented with a list of fields from the selected table. The first variable is set up the standard way, but the Obtained From parameter for the second variable is set to the first variable. In this example, the Select_Table variable is set to a data type of Table, and the Choose Field variable is set to a data type of Field with the "Obtained from" parameter set to the Select_ Table variable.

The value list for the second field is derived from the first field.

The user selects an existing table from the Data Type Examples dialog box. The input box for the second variable then shows all the fields in that table.

Only the field names from the selected table are listed as valid inputs.

The accompanying table from ArcGIS Desktop Help shows which data types have a valid "Obtained from data type" pairing.

Input data type	Obtained from data type	Description
Field or SQL Expression	Table	The table containing the fields
INFO Item or INFO Expression	INFO Table	The INFO table containing the items
Coverage Feature Class	Coverage	The coverage containing features
Area Units or Linear Units	GeoDataset	A geographic dataset used to determine the default units
Coordinate System	Workspace	A workspace used to determine the default coordinate system
Network Analyst Hierarchy Settings	Network Dataset	The network dataset containing hierarchy information.
Geostatistical Value Table	Geostatistical Layer	The analysis layer containing tables

The "Obtained from data type" settings can help simplify the user interface.

In the following exercise, the manager of a regional data warehouse for police department information has heard about your newfound knowledge of scripting and would like a script that lists the characteristics of shapefiles the warehouse gets from member cities each month. You'll need to design a model that gives the user a navigation dialog box to find only shapefiles. When a shapefile is selected, a printout of the data type and catalog path should be generated. In the exercise, you'll write a script that accepts an input, and then use the Describe command to list the required properties. When the script is added to your toolbox, you'll use the data type settings to restrict the input to only shapefiles.

Before you begin the exercise, examine the steps needed to complete the task:

- Create a new Python script.
- Write lines of code into the script editor.
- Include the Describe method in the script.
- Save the Python script.
- Add the script to a toolbox.
- Run and test the script.
- Examine the output of the Describe method.

Exercise 3e

1   Start the IDLE interface and open a new script window. Create a documentation section that contains the name of the script, its purpose, its arguments, and its creator information. Name the new script **Describe_SHP.py**.

```
7⁄ *Describe_SHP.py - C:\ESRIPress\GTKModelbuilder\MyAnswers\Describe_SHP.py

File  Edit  Format  Run  Options  Windows  Help

################################################################
# Describe_SHP.py
#
# This script will use a data type constraint and allow the
# user to select a shapefile.
# The outputs will be a print of the data type,
# and the catalog path.
#
# Arguments:
# 0 - Input Shapefile name (required)
#
# Your Name
# Month Year
# ArcGIS 10
#
################################################################
```

2   Next, add the necessary code to load the ArcPy module, as shown in the accompanying graphic.

```
# Load the arcgisscripting module
import sys, os, arcpy
```

3   Now add code to accept a single input.

```
# Get the input from the model
InputSHP = arcpy.GetParameterAsText (0)
```

4 Finally, add a set of print statements to invoke the Describe method. Check the Help page for the Describe method for more information.

```
# Create a describe object
#
descSHP = arcpy.Describe (InputSHP)

# Print shapefile properties
#
arcpy.AddWarning (descSHP.ShapeType)
arcpy.AddWarning (descSHP.CatalogPath)
```
Ln: 13 Col: 12

5 Save the script as Describe_SHP.py in your MyAnswers folder and close the IDLE window.

6 Start ArcMap and open EX03E.mxd. Copy the toolbox called Chapter 3e into your MyAnswers folder. Right-click the toolbox, and then click Add Script.

7 Type a script name of **DescribeSHP**, a label of **Describe Shapefiles**, and a description of **This model will print the data type and catalog path for each selected shapefile.** Select the check box to store relative path names, and then click Next.

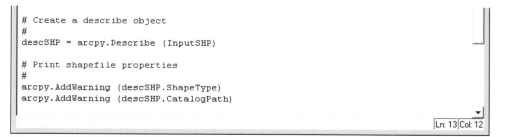

8 Use the Browse button next to the Script File text box to set the script file to the Describe_SHP.py file you just created. Click Next.

9 There is only one variable to set up. Type **Select Shapefile** for the display name and set the data type to Shapefile. Click Finish.

10 Double-click the Describe Shapefiles script to run it. This doesn't have to be in a model since there are no outputs to handle. Use the Browse button and navigate to the Data/RFDA Shapefiles folder. Select the BedfordCalls.shp file. Notice that only shapefiles can be selected. Click Add and then click OK.

11 In the Results window, note how much information is shown about the shapefile's parameters, data type, and catalog path. After reading the messages, close the Results window.

```
Results
├─ ⚠ Describe shapefiles [131844_07232010]
│  ├─ ◇ Inputs
│  │  └─ 💠 Select Shapefile: BedfordCalls.shp
│  ├─ 📋 Environments
│  │  ├─ 🗂 Precision For New Coverages: SINGLE
│  │  ├─ 🗂 Auto Commit: 1000
│  │  ├─ 🗂 Scratch Workspace: C:\Users\David\Documents\ArcGIS\Default1.gdb
│  │  ├─ 🗂 Minimize memory use during analysis on terrains: false
│  │  ├─ 🗂 Compression: LZ77
│  │  ├─ 🗂 Coincident Points: MEAN
│  │  ├─ 🗂 Random number generator: 0 ACM599
│  │  ├─ 🗂 Raster Statistics: STATISTICS 1 1
│  │  ├─ 🗂 Level Of Comparison Between Projection Files: NONE
│  │  ├─ 🗂 Output has Z Values: Same As Input
│  │  ├─ 🗂 Maintain fully qualified field names: true
│  │  ├─ 🗂 Tile Size: 128 128
│  │  ├─ 🗂 Pyramid: PYRAMIDS -1 NEAREST DEFAULT 75
│  │  ├─ 🗂 Default TIN storage version: CURRENT
│  │  ├─ 🗂 Output Spatial Grid 1: 0
│  │  ├─ 🗂 Cell Size: MAXOF
│  │  ├─ 🗂 Output has M Values: Same As Input
│  │  ├─ 🗂 Output Spatial Grid 2: 0
│  │  ├─ 🗂 Output Spatial Grid 3: 0
│  │  ├─ 🗂 Maintain Spatial Index: false
│  │  ├─ 🗂 Current Workspace: C:\Users\David\Documents\ArcGIS\Default1.gdb
│  │  └─ 🗂 Precision For Derived Coverages: HIGHEST
│  └─ 💬 Messages
│     ├─ ⓘ Executing: DescribeSHP "C:\ESRIPress\GTKModelbuilder\Data\RFDA Shapefiles\BedfordCalls.shp"
│     ├─ 🕐 Start Time: Fri Jul 23 13:18:44 2010
│     ├─ ⓘ Running script DescribeSHP...
│     ├─ ⚠ Point
│     ├─ ⚠ C:\ESRIPress\GTKModelbuilder\Data\RFDA Shapefiles\BedfordCalls.shp
│     ├─ ⓘ Completed script DescribeSHP...
│     └─ 🕐 Succeeded at Fri Jul 23 13:18:44 2010 (Elapsed Time: 0.00 seconds)
```

This is a very simple script wherein the results are printed only in the Results window. However, the outputs from the Describe method could also be used in an if-then-else or an if-elif-else statement to route the model based on data type or to use the catalog path to store other data.

What you've learned so far

◆ How to add and configure a method in a Python script
◆ How to restrict the input type in a variable
◆ How to display information from a model inside the geoprocessing Results window

This section covers many of the flow of control tools, methods, and techniques. The examples are designed to single out one technique at a time to make it easier to learn. Once you master these techniques, stringing them together into a single model would produce quite a complex program.

Chapter 4

Working within the modeling environment

The models, scripts, and toolboxes that make up the components of the ArcGIS ModelBuilder application are created and manipulated in a modeling environment. It is important to be able to control these components so that you can keep them running successfully, share them with other users, modify them to perform other tasks, or integrate them with other processes. You may even need to transform a modeling project from the ModelBuilder format into another programming language such as Python or others.

Sharing models

Once complex models are created, either as geoprocessing tasks or shortcut tools, they can be shared with co-workers and other users. Before doing this, however, it is important to understand how models and their data references are saved.

One important consideration is to preserve the stored paths of the data. Data variables that are added to the model maintain a fixed path to their source data. Moving the model to another location prevents the data variable from finding its associated data. The solution is to keep the relative path of the model to the data the same when the model is moved. This is done by setting the model to store relative path names instead of direct path names. A direct path name looks like this:

C:\ESRIPress\GTKModelbuilder\MyData\Storage.gdb\InputFeatures

If the recipient of the model does not have the same exact file structure built on their computer, the model will fail. This is very restrictive and does not allow much flexibility for others to use a shared model. A relative path name looks like this:

...\Storage.gdb\InputFeatures

This format tells the model to look in the folder where it has been copied and find the correct preceding system path. Of course, the data has to be copied along with the model for the relative path names to remain intact. Another option is to store the model in a geodatabase, ensuring that the file structure remains constant.

Setting the model to store relative path names is done in the model properties. A simple check box activates this feature.

Data links are not broken if the model is kept in relation to the data.

Another important consideration is where the model stores the data it creates. This is especially critical for intermediate data. Data storage can be controlled by setting the system environment paths for the current workspace location and the scratch workspace location. These locations then become available for variable substitution as system variables called %currentworkspace% and %scratchworkspace%.

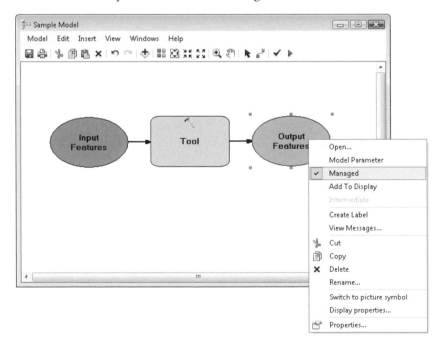

Setting these environment variables automatically activates them as system variables.

These variables can be made into model parameters and set by the user, or you can note in the model documentation that the user should set them. Once these variables are populated, the model can use them to store all the data it creates. Clicking the Managed option on the output variable's context menu allows all data outputs to be stored using the path shown in the current workspace environment setting.

Right-click an output data variable to access the context menu.

If an output variable is also set as intermediate in the context menu, it is stored using the path shown in the scratch workspace environment setting.

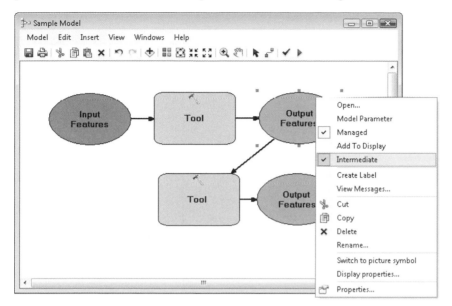

A managed intermediate output data variable is automatically directed to the scratch workspace.

An additional benefit to using the managed data settings is that the user does not have to type long or complex path names. By typing only the desired name of the output file, you ensure the file is saved, using the path stored in the current workspace environment settings.

When only the file name is entered, the model automatically uses the preset workspace path.

With extremely complex or time-consuming models, you may want to consider setting them to run as a background process. With this option activated, you start the model and fill in any required parameters, and then the model releases control of the processing environment while it continues to run in the background. You can continue working in ArcMap while the model completes its processes. When it is finished, a pop-up box will notify you. By default, models run in the foreground, which is useful in the writing and troubleshooting phases. But you can set them in the Model Properties dialog box to run as a background process if you'd like.

Sample Model Properties

General | Parameters | Environments | Help | Iteration

Name:

SampleModel

Label:

Sample Model

Description:

Storing relative path names is as simple as checking the box below.

Stylesheet:

☑ Store relative path names (instead of absolute paths)

☐ Always run in foreground

OK   Cancel   Apply

Clearing the "Always run in foreground" check box makes your model run in the background.

While the model is running, a status window is shown across the bottom of your map document. This lets you know that a background process is active.

|□ 回| ⟳ ‖ ◂   ‖   ▸ |◂

e Model...Sample M ⊙ 2406437.087 6993775.937 Feet   4.17 8.20 Inches

The model's name scrolls across this display until the background process is finished.

After the model has completed its run in the background, a pop-up box appears. The box simply reminds you to check on the results of your model and indicates whether the model ran successfully or encountered any errors.

⚠ **Sample Model**   ✕

A pop-up box appears briefly to indicate that a background process has finished running.

The pop-up window disappears within a few seconds, although you might want to click it and open the geoprocessing Results window. This window displays all the information about the model and any errors that it encountered.

The geoprocessing Results window displays messages about the geoprocessing tasks you run.

By setting these environment paths and other variable parameters, you make it easier not only to operate the model, but also to manage it when shared with other users.

In the following exercise, you have a model to do a simple buffer and overlay selection that you would like to share with others in your office. The problem is that when other users run it, the output files wind up in unknown places. In the exercise, you'll fix this problem by setting the system environment settings for both the current and scratch workspaces. Then you'll need to change the model's output variables to managed, which makes them automatically use the known system workspaces.

Before you begin the exercise, examine the steps needed to complete the task:

• Start editing the model.
• Set the system environment variables.
• Set the output variable parameters to use the system workspaces.
• Observe the %scratchworkspace% and %currentworkspace% system variables.

Exercise 4a

1   Start ArcMap and open EX04A.mxd. Copy the Chapter 4a toolbox from the SampleModels folder to your MyAnswers folder. Right-click the Buffer Parcels model in the toolbox, and then click Edit.

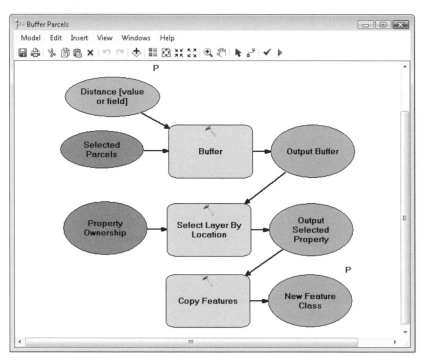

You'll first need to set the system environment variables for the current and scratch workspaces.

2   On the ModelBuilder menu bar, click the Model menu, and then click Model Properties.

3   Click the Environments tab, scroll down to the bottom of the list, and select the Workspace check box.

4 Click Values, and then click the chevron to expand the general settings.

ArcGIS has a default workspace set up in each user's folder, shown as Default.gdb. For this exercise, you are changing this to a geodatabase to be shared with the model.

5 Use the Browse buttons and set the current workspace to:

C:\ESRIPress\GTKModelbuilder\Data\CityOfOleander.mdb\Property Data

Set the scratch workspace to:

C:\ESRIPress\GTKModelbuilder\Data\CityOfOleander.mdb\Templates

Click OK and then OK again to save and close the Environment Settings dialog box. Remember that you are selecting a workspace, not a feature class, so you can select only geodatabases and feature datasets in this dialog box.

Environment Settings

⋀ Workspace
Current Workspace
C:\ESRIPress\GTKModelbuilder\Data\CityOfOleander.mdb\Property Data

Scratch Workspace
C:\ESRIPress\GTKModelbuilder\Data\CityOfOleander.mdb\Templates

OK   Cancel   Show Help >>

You can see from the model diagram that there are two variables set as model parameters. The final output feature class should be stored in the current workspace so that it can be located later. This is done by setting the variable as managed. In addition, the Output Buffer data variable should be stored as intermediate data in the scratch workspace. This is accomplished by setting the variable as managed, as well as setting it as intermediate.

6   Right-click the New Feature Class data variable and select Managed. This makes the model store the variable in the current workspace.

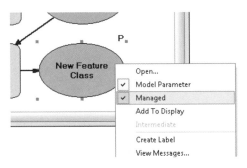

7   Finally, right-click the Output Buffer data variable and select both the Managed and Intermediate options. This makes the model store the variable in the scratch workspace, and because it is intermediate data, it can be overwritten by subsequent runs of the model. Save the model and close the model window.

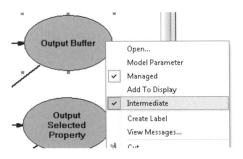

Setting the environment variables through the Model Properties dialog box makes them available for that model only. You can set them to be active for all geoprocessing operations by changing them in the Geoprocessing > Environments settings.

These environment variables can also be used in variable substitution as shown in earlier exercises by typing **%ScratchWorkspace%** or **%CurrentWorkspace%** into the path name.

8 Close the map document and exit ArcMap.

What you've learned so far

- How to set the scratch workspace and current workspace environment settings
- How to control the paths of a model's input and output data with the Managed option
- How to set and control the location of intermediate data
- How to use the system environment variables %currentworkspace% and %scratchworkspace%

Creating and sharing toolboxes

Models themselves cannot be shared with others. Rather, it is the toolbox that contains the model that is being shared, taking the model with it. A toolbox may contain many models, and you may not want to share all its contents, so it is a common practice to create a new toolbox and copy the model to be shared into that toolbox. Then that toolbox, or .tbx file, can be sent to others.

Toolboxes can be created in many different places. One of the most common is in the Toolboxes folder that ArcGIS creates and maintains. It can be found in the Catalog window of ArcMap or the Catalog tree in ArcCatalog, and it contains two subfolders. The My Toolboxes folder can store user-created toolboxes, while the System Toolboxes folder contains read-only ArcGIS toolboxes. This second set of tools cannot be modified, nor can a user create a new toolbox in this folder.

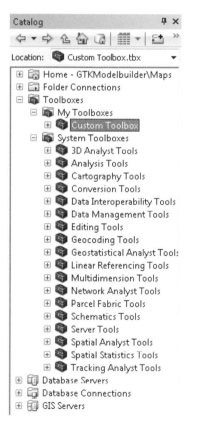

A location for custom toolboxes is provided in the default folder structure.

New toolboxes can be created in the provided My Toolboxes folder or in any folder except the System Toolboxes folder. Right-clicking a folder provides the New > Toolbox selection to begin the creation process. This allows a number of users to use private folders to store custom toolboxes they may not intend to share.

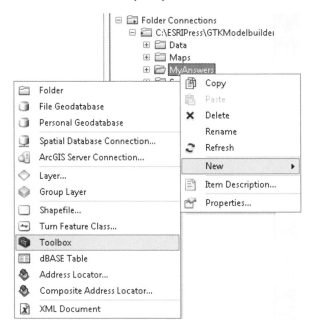

New toolboxes can be created in any folder except System Toolboxes.

Toolboxes can also be stored in a file geodatabase or personal geodatabase. They are created by using the context menu selection New > Toolbox. The toolbox structure is the same, but it is unique in that the model and its required data can be stored, and shared, in a single file. This involves sharing the entire contents of the toolbox as well as the entire contents of the geodatabase, which can be quite large, so due consideration must be made before using this technique. Also note that unlike other toolboxes, those that are created inside a geodatabase cannot have spaces or special characters in their name.

New toolboxes can also become part of a geodatabase.

Newly created toolboxes are symbolized by the toolbox icon 📦 and are visible in the system file viewer as a TBX file. These toolboxes can be copied and pasted into other folders. Geodatabase toolboxes, however, can only be copied into other geodatabases if you copy the entire geodatabase along with the toolbox. This may be a concern in determining how to share toolboxes.

These toolboxes are separate files and can be shared.

Toolboxes also have properties associated with them. The Properties dialog box is accessed by right-clicking a toolbox and then clicking Properties. The dialog box lets you set a name, a label, and a description, but of more interest is the option to have the toolbox refer to a compiled Help file (CHM). This file format can contain a rich set of documentation, including images, which could be used to describe all the models and scripts in your toolbox. A quick Web search can guide you to instructions on creating these files.

The toolbox properties can include a compiled Help file.

When users receive shared toolboxes, they can add them to their own folders by the method shown earlier. They can also use them in any geodatabase, and even publish them as a service in ArcGIS Server.

In the following exercise, you have created an interesting model that you think a colleague who works for another city might find useful, so you want to send it to him in an e-mail. The model's toolbox is stored a geodatabase that has many other models you don't wish to share. That's too much to send as an e-mail attachment anyway, so in the exercise, you'll make a new toolbox for this model. Then you can copy the model into the new toolbox and send the toolbox via e-mail to your colleague. He'll need to adapt it to his own datasets, but the processes are the important thing you are sharing.

Before you begin the exercise, examine the steps needed to complete the task:

- Create a new toolbox.
- Copy a model from one toolbox and paste it in another.
- Observe the resulting .tbx file properties.

Exercise 4b

1   Open EX04B.mxd in ArcCatalog. In the Catalog window, find the Data folder and expand the OleanderFireDept geodatabase. There you will see the OFD_Analysis toolbox.

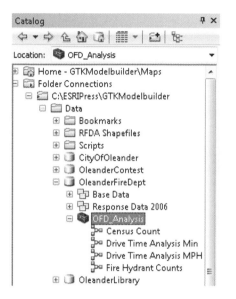

Notice that there are many models in this toolbox. The one you want to share is Census Count. You'll make a new toolbox for this model in your MyAnswers folder.

2   Right-click your MyAnswers folder, and then in the context menu, click New > Toolbox.

3   The new toolbox is created and shown in the Catalog window, with the file name in edit mode. Change it to **My Shared Tools** and press ENTER.

4   Go back to the OFD_Analysis toolbox and click it to display the models it contains. Right-click the Census Count model, and then click Copy.

5 Finally, right-click the new toolbox you created, and then click Paste.

The model is now the only one in the new toolbox, making it possible to share the toolbox and its single model with others.

6 Open Windows Explorer and navigate to the C:\ESRIPress\GTKModelbuilder\MyAnswers folder. In it, you will see the My Shared Tools TBX file that you can send in an e-mail to your colleague in another city.

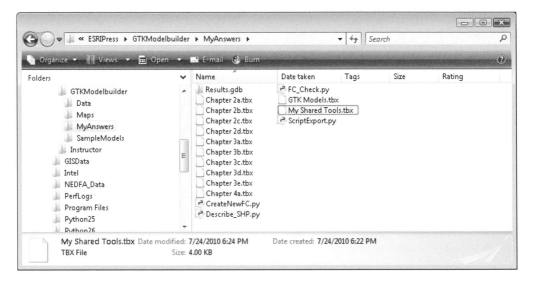

7 After examining the new toolbox, close Windows Explorer and exit ArcMap.

With the proper data storage techniques and good toolbox management, sharing models is pretty straightforward. If you experience problems, however, check ArcGIS Desktop Help and search the topic "Checklist for sharing tools." This document helps answer questions on sharing tools, models, and Python scripts, as well as dealing with multiuser access for ArcSDE and ArcGIS Server.

What you've learned so far

- ◆ How to create a new toolbox in a folder, in a geodatabase, or in the system default toolbox
- ◆ How to copy a model from one toolbox and paste it in another
- ◆ How to identify a toolbox file in Windows Explorer

Validating a model

Sometimes when models are shared or moved, or even when the input data location or structure is changed, some of the model elements may return to the Not Ready to Run state. This keeps the model from executing successfully, and when the user opens the model in the model window, some of the components will appear with a hollow, colorless symbol.

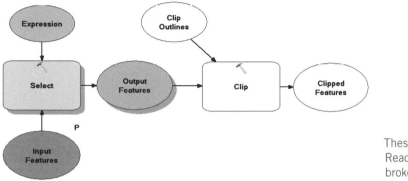

These elements in the Not Ready to Run state have broken data links.

To fix this problem, the user must reset the data locations for each of the data variables in the Not Ready to Run state and validate the entire model again. Validating the model also returns the components that have been run to the Ready to Run state.

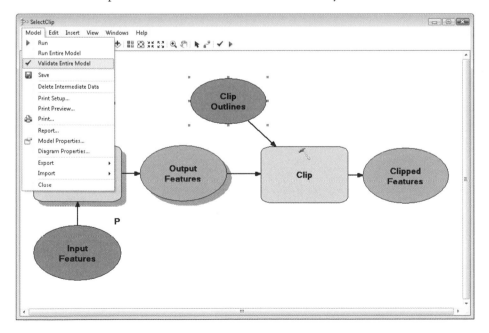

Validating the model corrects broken data links.

Once the model is validated, it can be saved and run under normal conditions. Models may also have problems with overwriting existing files after they have been moved. This may be because of a setting in the ArcGIS system environment that controls the intermediate files. Selecting the "Overwrite the outputs of geoprocessing options" check box in the Geoprocessing Options dialog box gives the model permission to overwrite existing files.

The geoprocessing options control many aspects of the tools used in ModelBuilder.

Another thing to try when a shared model won't run is to delete any existing intermediate data files. This is done rather easily from the Model menu. Once this intermediate data is deleted, the conflicts with the existing data are eliminated. Note that after the intermediate data is deleted, the model is automatically validated again.

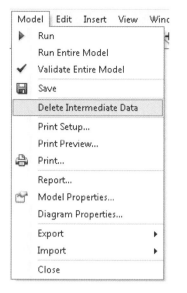

Sometimes intermediate data is not automatically deleted after a model has finished running, but this process clears it.

If any problems are experienced after a model is shared, these techniques should help to rectify the situation. Once fixed, the models should run on the new computers as successfully as they did at the source location.

In the following exercise, one of your colleagues has sent you a model, and it doesn't seem to be working. In the exercise, you will review the techniques for troubleshooting shared models and see if you can get it to work on your computer. You'll start by checking the setting for overwriting geoprocessing files, and then run through the other techniques for troubleshooting models.

Before you begin the exercise, examine the steps needed to complete the task:

- Set the system options for geoprocessing.
- Start editing the model.
- Use the Delete Intermediate Data command.
- Set the model parameters.
- Validate the model.
- Save and close the model.

Exercise 4c

1 Start ArcMap and open EX04C.mxd. Copy the Chapter 4c toolbox from the SampleModels folder to your MyAnswers folder.

2 On the ArcMap main menu bar, click Geoprocessing > Geoprocessing Options. Note the status of the selection "Overwrite the outputs of geoprocessing operations." Select the check box to activate this option, if necessary. Click OK to close the dialog box.

General

☑ Overwrite the outputs of geoprocessing operations

☑ Log geoprocessing operations to a log file

3 Next, start editing the Buffer Selected Parcels model. Notice that the Buffer tool has already been run, so you must return it to the Ready to Run state and delete the file it has created. On the ModelBuilder menu bar, click Model > Delete Intermediate Data.

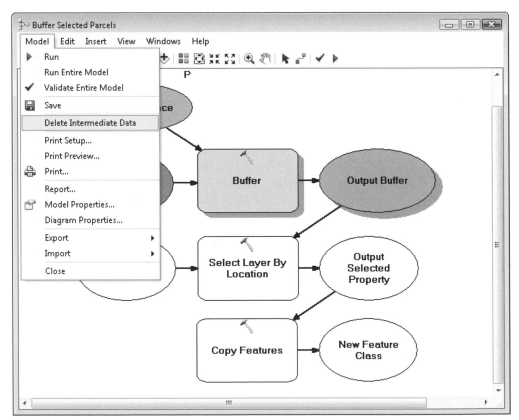

These two steps, overwriting outputs and deleting intermediate data, cure a lot of problems that models have, but this model still appears to have a problem. The model still has several tools in the Not Ready to Run state. First off, you'll try to get the hollow components into the Ready to Run state.

4 Double-click the Parcels data variable. In the Parcels dialog box, set the input data to Property Ownership. Click OK.

Doing this step should put all the hollow components into the Ready to Run state.

5 Now you can have the model check all the paths for both the input and output data variables by clicking Model > Validate Entire Model on the ModelBuilder menu bar or by clicking the Validate Entire Model button on the ModelBuilder toolbar.

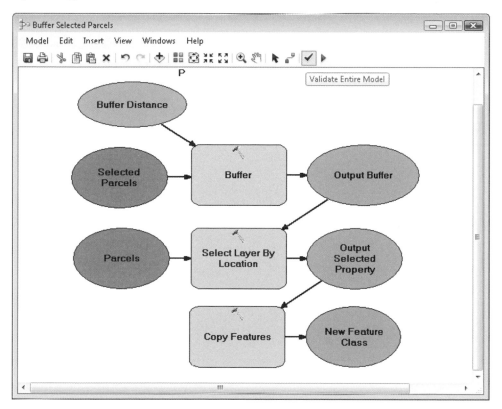

6 Finish the process by saving the model and closing the model window. You can run the model if you wish, but the objective is to get it in the Ready to Run state for your colleague.

These steps correct several errors in the model and return it to a usable state. If you receive a model with these types of errors, try these techniques to get the model running again. It may also be necessary to contact the author of the model for more information about the source data locations.

What you've learned so far

- How to set a model to overwrite existing output files
- How to use the Delete Intermediate Data command to clear previous results
- How to validate the paths and data of a model and prepare the model to run

Working within the modeling environment   183

Exporting a model to a Python script

In chapter 3, you learn how to write your own custom Python scripts from scratch, and these Python scripts are very easy to share with others. In some scenarios, it might even be easier to share a script than it is a model, and ArcGIS includes a process for exporting a model to a Python script.

When a model is open in the model window, the user can click Model > Export > To Python Script.

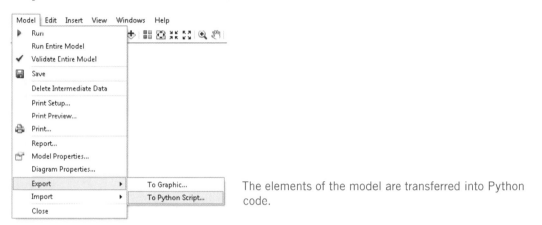

The elements of the model are transferred into Python code.

The output of this command is formatted as a Python script, which can then be modified by notations or other commands.

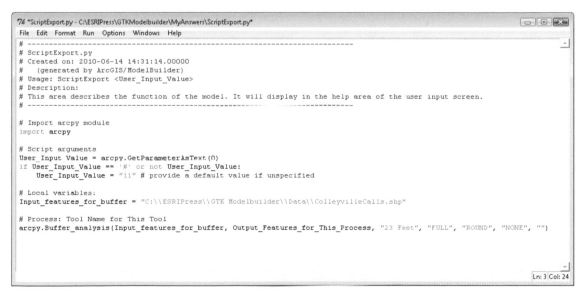

This script can be used by anyone using Python.

The script can then be shared with others in the same way that models are shared, except that the entire toolbox doesn't need to be copied. The recipient of the script may need to correct some paths in the script, or change a file name, but the code is transportable.

Another particularly useful application of this Export tool is to get a quick sample of code for a larger script you may be writing. For example, if you are writing a script that needs to create a new feature class, you can drop that tool into a model, export the model to a script, and then get a block of code you can incorporate into your script.

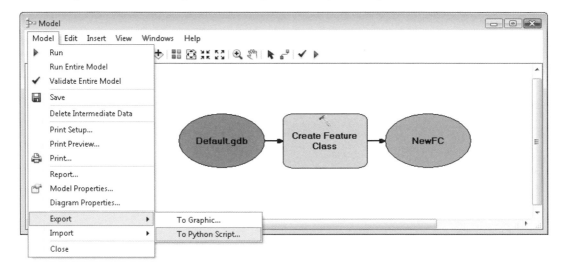

Python code for the single tool Create Feature Class is output.

You can open the script in the Python editor IDLE, which loads with ArcGIS, and copy and paste it into another script.

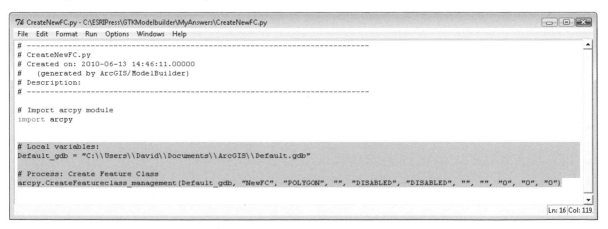

This snippet of code allows you to see the tool's syntax.

Examining the code syntax of a single tool can help you to become a better Python programmer and guide you in setting up more complex routines in your code. Remember also that Python scripts can be included in a model, so you can use the Export tool to break the steps of your model into individual components. As each component is completed and tested, it can be added to a more complex model.

In the following exercise, the GIS manager in the neighboring city of Grapevine is impressed with the Create Oleander FC model in which you added a Python code in exercise 3d. She's writing a similar process as a script and would like your model in the form of Python code she can incorporate in her project. In this exercise, you'll export the model to a script to give her the code she is requesting. Then you can add notes to the script so she can better understand the processes in your script. Even though she is a better Python programmer, she won't know that you didn't write this code from scratch!

Before you begin the exercise, examine the steps needed to complete the task:

- Open the model for editing.
- Export the model to a Python script.
- Add notations to the code.

Exercise 4d

1   Start ArcMap and open EX04D.mxd. Navigate to the SampleModels folder and find the Chapter 4d toolbox. Open the toolbox and start editing the Create Oleander FC model.

2   Click Model > Export > To Python Script to start the export process.

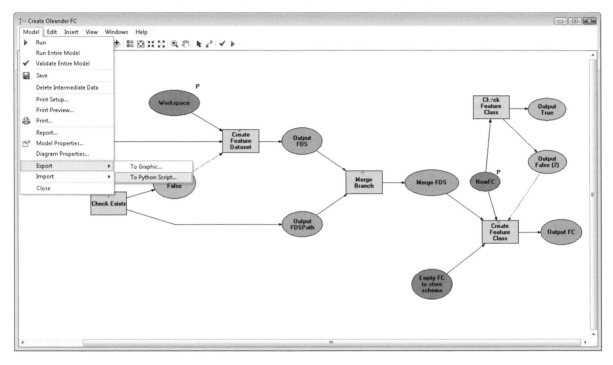

3   Save the script as **CreateNewFC** in your MyAnswers folder. Close the model window and exit ArcMap.

4 Start the IDLE editing window and open the CreateNewFC.py script you just created. Move through the script and add notes as necessary to explain the processes it contains.

```
# --------------------------------------------------------------
# CreateNewFC.py
# Author: David W. Allen, GISP
# Created on: 2010-07-24 23:49:58.00000
#   (generated by ArcGIS/ModelBuilder)
# Usage: CreateNewFC <Workspace> <New_FDS_Name> <NewFC>
# Description:
# Create a new Feature Class with the City of Oleander parameters and
# spatial reference pre-set
# --------------------------------------------------------------

# Import arcpy module
import arcpy

# Load required toolboxes
arcpy.ImportToolbox("C:/ESRIPress/GTKModelbuilder/MyAnswers/Chapter 3d.tbx")
arcpy.ImportToolbox("C:/ESRIPress/GTKModelbuilder/SampleModels/Chapter 3c.tbx")

# Script arguments
Workspace = arcpy.GetParameterAsText(0)
if Workspace == '#' or not Workspace:
    Workspace = "C:\\ESRIPress\\GTKModelbuilder\\MyAnswers\\Results.gdb" # provi

New_FDS_Name = arcpy.GetParameterAsText(1)
if New_FDS_Name == '#' or not New_FDS_Name:
    New_FDS_Name = "New FDS Name" # provide a default value if unspecified

NewFC = arcpy.GetParameterAsText(2)
if NewFC == '#' or not NewFC:
    NewFC = "NewFC" # provide a default value if unspecified

# Local variables:
Output_FDS = Workspace
Merge_FDS = Output_FDS
Output_FC = Merge_FDS
Empty_FC_to_store_schema = "Empty FC to store schema"
Output_True = NewFC
Output_False__2_ = NewFC
Output_FDSPath = New_FDS_Name
Output_False = New_FDS_Name
```

5 When you have added sufficient notations, save and close the script. The script can be run just like the model, and it calls the geoprocessing tools that are necessary to perform the same operations as the model. This script is also ready to send out, and you don't need to tell the GIS manager in Grapevine that you didn't write this script from scratch.

By looking through the script, you can get an idea of how Python code is written for each of the tools that is in the model. Similar to the model, the script prompts the user for certain input values, and then calls the ArcGIS tools necessary to perform the tasks.

What you've learned so far

- How to export a model to a Python script
- How to identify model components and tools when written as Python code
- How to document a script to be shared with others

Chapter 5

Using multiple inputs

A powerful aspect of the ArcGIS ModelBuilder application is the ability to work with many inputs in a single model. This may include multiple datasets of the same type that need the same type of processing, or it may involve a collection of different dataset types that the model will have to scrutinize, and then be able to apply the correct processes to those datasets. Whatever is the case, ModelBuilder provides many ways to work with multiple inputs.

Batch processing with user input

Models provide a great way to automate a task by letting the user build a diagram containing the necessary processes, but so far in earlier chapters, the tasks have all been run on one dataset at a time. The user runs the model once for each dataset used as input. However, if multiple datasets require the same processing tasks, the model can still be run just once with multiple inputs provided by use of a batch or series variable. The user then specifies more than one input file or parameter value to create multiple outputs.

This function exists by default for all ArcGIS geoprocessing tools and can be performed outside a model. Right-clicking any geoprocessing tool and then clicking Batch opens the batch grid control. By populating each row with all the necessary input data, the tool runs once for each valid row of data. In the following example, the staff of the Oleander Library needs to search historical data to locate patrons with more than 50 transactions in a single year. Each row contains all the parameters necessary to run the Select tool, including optional parameters that may not be used. Notice that the expression in the accompanying diagram is the same for each row.

The batch grid control is available for any geoprocessing tool.

The batch grid will collect the parameters for a tool process in columns and rows. Each column represents an input parameter, and each row represents one set of values. The underlying tool runs as if its parameters are defined as a list of values. You can accomplish the same thing in a model by creating list variables. Note that input list variables cue a process to execute once per value in the list. This should not be confused with the Multivalue option, which supplies multiple values to a single process that accepts multiple inputs.

Input Features Properties

General | Data Type | Layer Symbology

This variable contains:

○ A single value

◉ A list of values

Feedback

Feedback Variable:

[OK] [Cancel] [Apply]

This variable is set to accept a list of input values.

When a variable is set to allow multiple inputs, its symbol changes to show several ovals. In this example of using the Select tool inside a model, the properties of the input variable are changed to allow a list of values. Setting the variable as a model parameter opens the batch grid input screen when the model runs. Note that the output variable is automatically changed to allow for multiple values.

The stacked ovals represent multiple inputs.

The batch grid, as shown previously, contains all the parameters of the tool. When used inside a model, the batch grid contains only the variables marked as model parameters. Stand-alone variables can also be included in the batch grid. The accompanying example model now contains a stand-alone variable, which is marked as a list variable and a model parameter. The Expression variable, however, is not marked as a list variable, which means that the same expression is used for each input value.

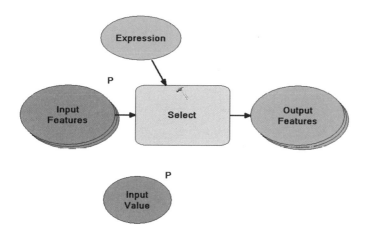

The Select tool runs once for each value in the Input Features variable, using the same expression each time.

Notice in the results from running the example model that only the variables that are marked both as multiple values and as model parameters appear in the batch grid. This gives the programmer maximum control over what inputs are shown to the user, an advantage over running the tool in batch mode from the Catalog window. The output variable is not marked as a model parameter, so each output is given a name by default.

Variables set to accept a list of inputs are shown in the column headers in the batch grid control.

The multiple outputs perform exactly the same as a single output in terms of being set as intermediate data, adding to the display, or being used as input for a second process. When the connections are made in the model, the output variable for the connected tool is automatically marked as having multiple values.

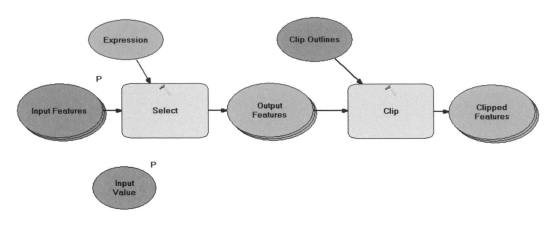

Both tools shown act upon each item in the list.

Variable substitution is also honored in a multiple-value process. The stand-alone variable is displayed in the batch grid and a value is given for each set of parameters. The value is substituted exactly as if the model were being run for only a single set of data.

In the following exercise, several cities near Oleander have grouped together to form a mutual-aid organization for their fire departments. The Regional Fire Department Association (RFDA) provides specialty equipment such as a hazmat team, a technical rescue team, a bomb squad, and an emergency GIS response team. Once a year, all the data for calls for service from the member cities is compiled by the GIS team for various types of analysis across the district.

The Data\RFDA Shapefiles folder has shapefiles from the cities of Oleander, Hurst, Bedford, Colleyville, Keller, and Southlake—all the RFDA members. The GIS team needs to get a total of how many responses each file contains for a report to the state.

In the exercise, you'll write a model to perform this task. It should be easy to write. The model needs to use the Get Count tool to count the features. The difficulty is that there are many files to analyze, making it time consuming to run the model multiple times and necessary to keep track of which files have been checked to avoid duplication. The solution is to make the input variable accept multiple values. That way, the model can process all the files at once.

Tip: The instructions for the basic steps of creating new models, finding tools with the Search window, dragging them into the model, and connecting them are no longer described in the following exercises. If you need help with these steps, review the exercises in earlier chapters.

Before you begin the exercise, examine the steps needed to complete the task:

- Create a new model.
- Find a tool and add it to the model.
- Expose a tool variable as a model variable and set it as a model parameter.
- Change the input variable to a list variable.
- Save the model.
- Run the model and work with the multiple-value input dialog box.

Exercise 5a

1 Start ArcMap and open EX05A.mxd. Create a new toolbox called **Chapter 5a** in your MyAnswers folder.

2 Right-click the Chapter 5a toolbox and create a new model. Name it **CountFireCalls** with a label of **Count Fire Calls** and a description of **Get a count of the features in each shapefile. Note: Uses multiple inputs.** Save the model.

3 Find the Get Count tool and drag it into the model.

4   Expose the input rows of the Get Count tool as a variable and make it a model parameter. Set its value to C:\ESRIPress\GTKModelbuilder\Data\RFDA Shapefiles\ BedfordCalls.shp.

This is basically the way the model looks if you were going to process the files one by one—a single input resulting in a single output.

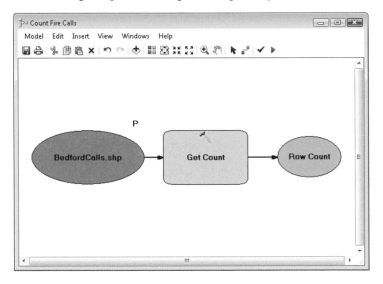

5   Right-click the input variable (labeled BedfordCalls.shp) and open the BedfordCalls.shp Properties dialog box. Click the General tab and select "A list of values." Click OK.

6 Notice that the symbol for the Input Rows variable has changed to show that it is working with multiple values. The symbol for the Row Count variable has also changed. Save and close the model.

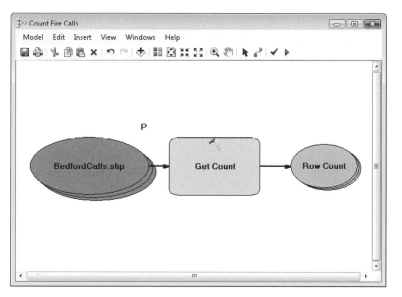

7 In the Chapter 5a toolbox, double-click the Count Fire Calls model to run it. The model opens the batch grid control window. Notice that there is only one input row, along with the single input cell that has the default value you specified.

8 To add more values, click the Add Row button ✚. A new empty row is added.

9 Right-click in the new row, and then click Browse (or double-click in the empty row). Navigate to the Data\RFDA Shapefiles folder and select ColleyvilleCalls.shp.

10 Click Add and note that this file is added to the Count Fire Halls dialog box.

This step adds one more file, but there are many more to add. The batch grid needs to have another data entry row.

11 Click Add Row in the Count Fire Halls dialog box. Another data entry row is added. Right-click in the row, click Browse as before, and add the HurstCalls shapefile. One by one, repeat the steps, adding the files KellerCalls.shp, OleanderCalls.shp, and SouthlakeCalls.shp. **Tip:** To make this a little easier, add four new rows to the batch grid you are creating. Find the source folder (RFDA Shapefiles) in the Catalog window, and then drag each file into the grid on a different row.

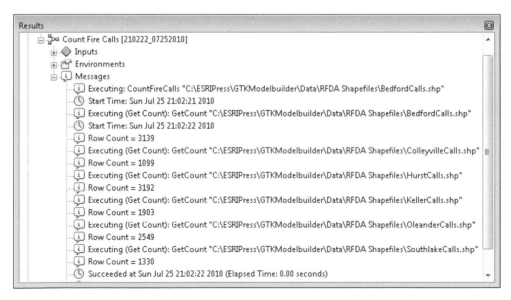

12 Click OK in the Count Fire Halls dialog box to run the model using a batch grid. In the Results window, you can see the feature count for each input file. After reviewing these numbers, close the Results window.

This is a pretty simple model, but it illustrates the power of batch processing.

The data collected for the RFDA is used in this model for a summary operation, but it may also be useful in locating areas with poor fire protection. Using spatial statistics tools, it is possible to look at the probability of future calls for service and determine both the optimal placement for fire stations as well as the optimal routing to respond to fire calls.

What you've learned so far

♦ How to set up a variable to accept a list of values as input to a model
♦ How to use the batch grid to enter multiple values into a list variable

Batch processing using lists

Another way to input the files for a batch process is to use a list or a table. The list can come from any format that ArcGIS can read, including database tables, delimited text files, and even spreadsheets. The variables in the model are set to accept multiple files as user input, but the user can also populate the batch grid by selecting a single table.

The batch grid can be set to read the values from a table.

The first line of the table should contain a field name corresponding to the number of inputs in the batch grid. The subsequent lines should contain the values that will be read into the grid. As with other batch inputs, each record in the table becomes a row in the batch grid and is used when the model runs.

In the following exercise, the library manager also has some files that the staff would like to get feature counts for. The data represents a summary of patron activity in a year. Each feature in the feature class represents one patron, and the library wants to know how many patrons used its services in each of the years for which the library has data. One difference from the fire department model is that the library has a table listing all the files it wants to process. In the exercise, you'll create a model that can use a list to process these files.

Before you begin the exercise, examine the steps needed to complete the task:

- Create a new model.
- Add geoprocessing tools to the model.
- Expose an input parameter as a model variable and make it a list variable.
- Set the list variable to accept input from a list.
- Set field name parameters.
- Save and run the model.
- Examine the results in the Results window.

Exercise 5b

1 Start ArcMap, if necessary, and open EX05B.mxd. Create a new toolbox in your MyAnswers folder called **Chapter 5b** and create a new model in it called **PatronCount**. You may invent your own label and description, and then save the model.

2 Drag the Get Count tool into the model and expose the Input Rows variable.

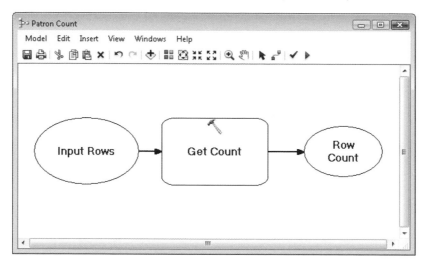

3 Right-click the Input Rows variable and open its properties. Set the variable to use a list of values.

4 Double-click the Input Rows variable. Right-click in the input box and select "Insert values from a table."

5   Navigate to the Data folder and select Patrons.txt. This is a comma-delimited text file that contains the path and file name for each feature class that needs to be processed. Cllck Add.

6   The Select Field dialog box asks which field contains the file names. Click FilePath, and then click OK.

7 The model reads all the file names from the table and adds them to the batch grid. Click OK.

Input Rows

	Input Rows
1	C:\ESRIPress\GTKModelbuilder\Data\OleanderLibrary.mdb\Year 1996\PatronActivity1996
2	C:\ESRIPress\GTKModelbuilder\Data\OleanderLibrary.mdb\Year 1997\PatronActivity1997
3	C:\ESRIPress\GTKModelbuilder\Data\OleanderLibrary.mdb\Year 1998\PatronActivity1998
4	C:\ESRIPress\GTKModelbuilder\Data\OleanderLibrary.mdb\Year 1999\PatronActivity1999
5	C:\ESRIPress\GTKModelbuilder\Data\OleanderLibrary.mdb\Year 2000\PatronActivity2000
6	C:\ESRIPress\GTKModelbuilder\Data\OleanderLibrary.mdb\Year 2001\PatronActivity2001
7	C:\ESRIPress\GTKModelbuilder\Data\OleanderLibrary.mdb\Year 2002\PatronActivity2002
8	C:\ESRIPress\GTKModelbuilder\Data\OleanderLibrary.mdb\Year 2003\PatronActivity2003
9	C:\ESRIPress\GTKModelbuilder\Data\OleanderLibrary.mdb\Year 2004\PatronActivity2004

OK   Cancel   Apply   Show Help >>

8 The symbols of the variables change to show that they are processing multiple values. Save the model. Since this model has no model parameters, it can be run from the model window. Click the Run button on the ModelBuilder toolbar.

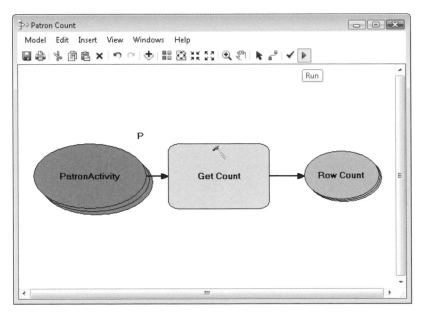

9 The Results window shows the results of the Get Count tool. Close the Results window. Then save and close the model. Exit ArcMap.

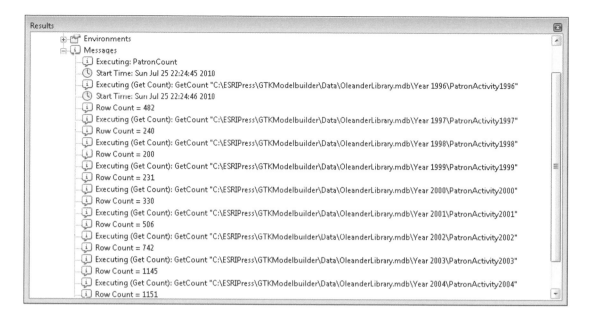

The location of library patrons might also be used to find the optimal locations for library branches. When combined with apartment locations and street centerlines, it could be used to route a bookmobile to the more populous areas of town.

What you've learned so far

♦ How to set the batch grid of a list variable to accept values from a file

Chapter 6

Using model iterations

Running a model can be a powerful way to perform a task, but sometimes running a model just once won't do the trick—it has to be run multiple times. This may be because you are using different input values on the same datasets or because you have many datasets to process. With the model iteration tools, or iterators ↺, you can control how many times the model runs based on a number of parameters. As you can see in this chapter, there are many variations on model iterations, including the use of iterators, and when these techniques are combined with other model controls such as Python scripting, the possibilities expand greatly.

Running a model a fixed number of times

Model iterations involve running a model a controlled number of times. This is generally based on the actions or results of running the model and is an important component of making your models more reactive to the processes they perform. The simplest of the model iteration types, and the most direct, is simply stating the number of times a model should run.

The programmer enters a number in the model properties that defines how many iterations the model should perform. The model then runs this number of times and stops.

5

6

7

Model Properties	? ✖

General | Parameters | Environments | Help | Iteration

● Run the model the following number of times:

> 7

○ Get the iteration count from this variable:

> ▼

○ Run the model until this variable is false:

> ▼

Maximum number of interations:

> 1

OK   Cancel   Apply

This model will run seven times, and then stop.

When this method is used, it is important to set the final output variable to create a new file name for each iteration. Otherwise, the model's output is overwritten after each iteration, and no usable data is produced. The system variable %n% can be included in the output name, which adds a number corresponding to the iteration number (starting at 0).

The system variable keeps the output file names unique.

An example of this method is modeling the spread of a wildfire. A polygon can be drawn to show the current fire line. Then a model can be used to predict the expected expansion of the fire by some percentage for time frame 1. The new expanded polygon is then expanded by the same percentage for time frame 2. Basically, the same process is repeated for the same dataset, using the newly created feature for each time frame. The programmer will preset how many iterations to run, which may be the number of time periods that the predictive data will remain valid.

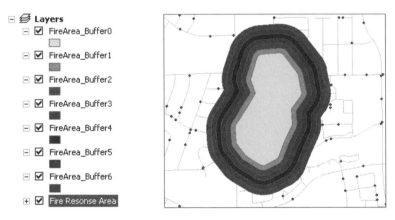

Each buffer is placed in a new file with the incremental numbering starting at 0.

In the following exercise, the fire chief has a plan for you to work on. He's been reading *ArcNews* and has an idea for an analysis he'd like to try. It tests the idea that a call for service to the fire department may generate more calls for service within the immediate area. He wants to pick a call at random and see how many calls happened within 500 feet of it. Then he'll take that new selected set and see how many calls happened within 500 feet of those calls. He'll repeat the process several times. Once all the numbers have been gathered, he'll put them into a spreadsheet and finish reading the article to see what to do next. Your only responsibility is to provide the count data.

In the exercise, you'll have the user select a single call for service before running the model. When the model runs, it uses a Select By Location process with a search distance to find all the calls for service within 500 feet. The selection type should be set to add the new features to the selected set rather than having the model clear the selected features and create a new selection. You will need to make the model repeat this process 11 times, for a total distance of 5,500 feet (a little over a mile).

Before you begin the exercise, examine the steps needed to complete the task:

- Create a new model.
- Add tools to the model and configure them.
- Set the model iteration parameters.
- Save and run the model.
- Examine the results in the Results window.

Exercise 6a

1 Start ArcMap and open EX06A.mxd. In your MyAnswers folder, create a new toolbox called **Chapter 6a**. Create a new model called **CFSSpread** with a label of **Calls for service spread** and a description of **Analyze the spread of service calls for the Fire Department.**

2 Search for the Select Layer By Location tool, and then add it to the model.

3 Double-click the tool to open its parameters. Set the input feature layer to Response2006 and the selecting features to Response2006. Next, change the relationship to WITHIN_A_DISTANCE and set the search distance to **500** feet. Finally, change the selection type to ADD_TO_SELECTION and click OK.

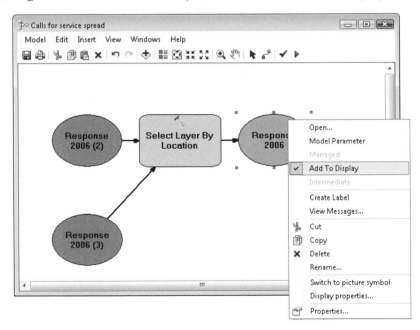

4 Right-click the Select tool output, and then click Add To Display.

5 Open the Calls for service spread Properties dialog box and click the Iteration tab. Set the number of iterations to **11**. Click OK.

```
Calls for service spread Properties                    [?] [X]

 General | Parameters | Environments | Help | Iteration

    ⦿ Run the model the following number of times:

       ┌──────────────────────┐
       │ 11                   │
       └──────────────────────┘

    ○ Get the iteration count from this variable:

       ┌────────────────────────────────────┬──┐
       │                                     │ ▾│
       └────────────────────────────────────┴──┘

    ○ Run the model until this variable is false:

       ┌────────────────────────────────────┬──┐
       │                                     │ ▾│
       └────────────────────────────────────┴──┘

       Maximum number of interations:

       ┌──────────────────────┐
       │ 1                    │
       └──────────────────────┘



                    [   OK   ]  [ Cancel ]  [ Apply ]
```

By setting a finite number of iterations, you are controlling the overall search distance so that it equals 5,500 feet. After 11 iterations, the model will stop.

Note that the model parameters also accept the number of iterations to perform from a variable. It is possible to add a stand-alone variable to the model, make it a model parameter, and use it as the iteration control with variable substitution.

6 The last step is to add the Get Count tool. Find it and drag it into the model. Connect the output of the Selection tool to the Get Count tool as the input rows. Save and close the model.

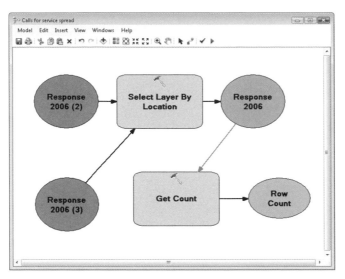

7 Make sure that a feature in the Response2006 layer is selected, and then double-click the model to run it. Note that it has no user input parameters, so click OK to continue. The model goes through 11 iterations and produces a count of the features at each iteration. Refresh the map to see the selected features.

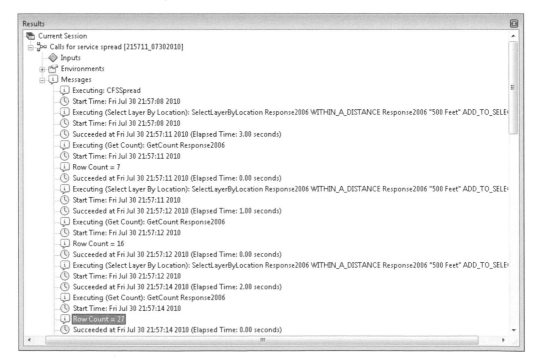

The numbers can be recorded for future use. If you like, try selecting a different point and running the model again. The fire chief can run this model many more times and hopefully be able to compile some useful data from it.

This model performs several ring selections to get a table of information. A more complex version of the model could quantify the spatial relationship of buffers created around adjacent call locations. This would show how one call for service could affect the probability of a call for service originating from other nearby sites.

What you've learned so far

+ How to set a finite number of iterations a model will run

5

6

7

Iterating in a model by Boolean expression

Another method of controlling the iteration of a model is to set a variable as the ending condition and have the model check it. The condition variable should be a Boolean value, or a value that can be detected as either true or false. Note that with integer fields, values of 0 or less evaluate to false, while positive values evaluate to true. In the model properties, the output variable containing the Boolean value is set as the iteration variable. When this is activated, it is also recommended that a maximum number of allowed iterations be set. If the condition statement in the model is incorrect, or if there is some anomaly in the data, the iteration control prevents the model from running endlessly.

One method of implementing this type of iteration is to have the model produce a value that represents the status of the model, and then check it with an expression in the Calculate Value tool. The model stops when the script returns a value of false. This is the equivalent of the Do While, Do Until, and For loops that are available in other programming languages.

```
Calculate Value                                        [x]

Expression
getVal(%MyVariable%)
Code Block (optional)
def getVal(intValue):
  if intValue > 0:
    return "TRUE"
  else:
    return "FALSE"


           OK      Cancel      Apply     Show Help >>
```

An iteration can be established in a Python script.

The Calculate Value tool should be made a precondition of the final process so that the evaluation takes place as the last process. This isn't necessary if list or series variables are used, because they have the power to stop the model once all the data is processed.

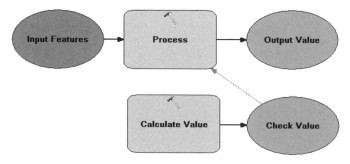

A Python script sets a precondition on the model's process.

The script can be very simple, or it can be developed into a complex process depending on the programmer's skills. The example of a model predicting the growth of a fire could have an iteration script to check the size of the polygons being generated. A maximum value could be set, and the model would then stop when that value is reached.

In the following exercise, the city planner wants to check the population density of Oleander against a hypothetical set of data to see how much it varies from the norm. Oleander has a population of 70,000 and an area of 11 square miles. To get a hypothetical test set, she must first select a census tract and check its population value. Then she'll need a model to select all the census tracts that share a boundary and add up their population values. If the population total for the selected census tracts comes in under 70,000, the model will go through another iteration. When the total population exceeds 70,000, the model will stop, and the city planner can retrieve the total population figure and the area of the census tracts it took to generate that number. Compiling many sets of this data allows the city planner to compare Oleander's population density to other cities in the region and could support some of her theories on urban sprawl.

In the exercise, you will automate the process of compiling a hypothetical set of data based on a user-selected census tract.

Before you begin the exercise, examine the steps needed to complete the task:

- Create a new model.
- Add and configure the Summary Statistics tool.
- Set the model iteration parameters to read a value from the data.
- Save and run the model.
- Examine the results in the map display.
- Examine the resulting summary statistics table.

Exercise 6b

1 Start ArcMap, if necessary, and open EX06B.mxd. The map document contains a set of census data for the county.

2 In your MyAnswers folder, create a new toolbox called **Chapter 6b**. Create a new model with the name **CensusCount**, a label of **Census Count**, and an appropriate description.

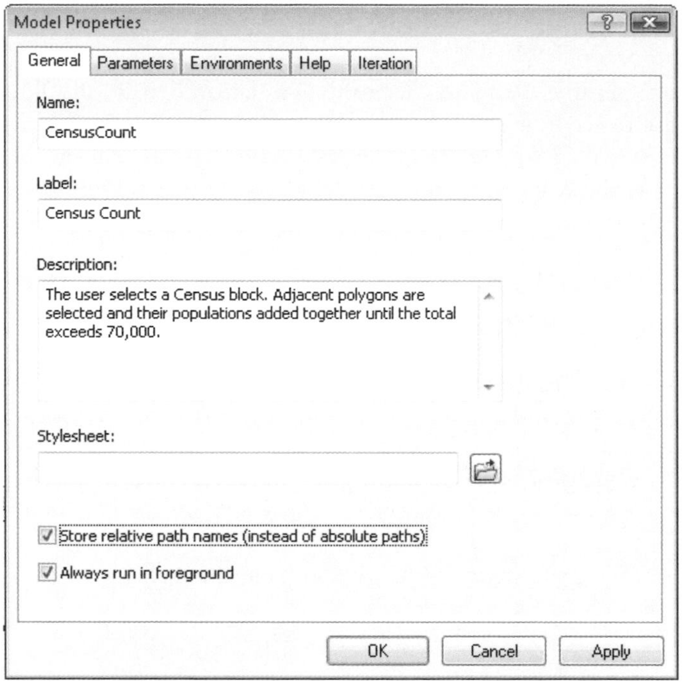

Model Properties

General | Parameters | Environments | Help | Iteration

Name:

CensusCount

Label:

Census Count

Description:

The user selects a Census block. Adjacent polygons are selected and their populations added together until the total exceeds 70,000.

Stylesheet:

☑ Store relative path names (instead of absolute paths)
☑ Always run in foreground

OK   Cancel   Apply

3 Find the Select Layer By Location tool and add it to your model. Set both the input feature layer and the selecting features to County Census. Change the relationship to BOUNDARY_TOUCHES and the selection type to ADD_TO_SELECTION. Click OK.

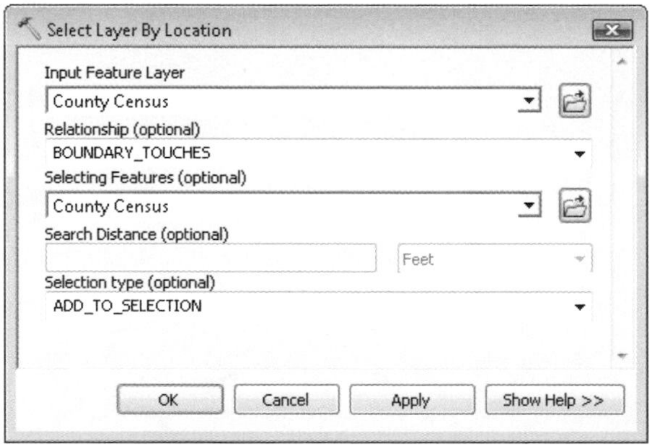

Select Layer By Location

Input Feature Layer

County Census

Relationship (optional)

BOUNDARY_TOUCHES

Selecting Features (optional)

County Census

Search Distance (optional)

[   ] Feet

Selection type (optional)

ADD_TO_SELECTION

OK   Cancel   Apply   Show Help >>

4 Now find the Summary Statistics tool and drag it into the model.

5 Use the Connect tool to connect the County Census output variable (shown in green) to the Summary Statistics tool as the input table.

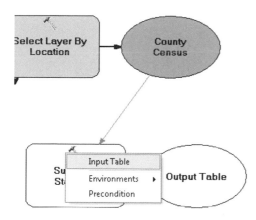

6   Open the parameters of the Summary Statistics tool. In the Statistics Field column, select TOTAL_POP. Next, click in the Statistic Type column and select SUM. Then go to the Statistics Field column again and select SqMiles. Set its statistic type to SUM. Click OK.

The output of this process is a summary table of the total population count and the total area in square miles each time the model iterates.

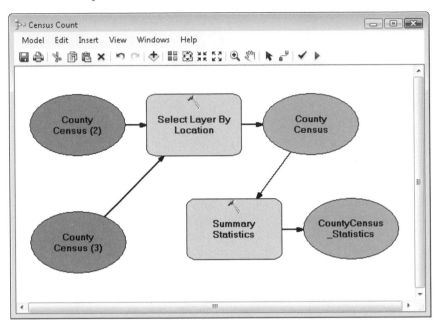

7   Right-click the CountyCensus_Statistics variable, and then click Add To Display.

The Table Select tool selects values from the summary table. It also has an option to set an expression, or query, that is applied when the tool is run. For this project, the expression checks to see whether the value is less than 70,000, the threshold at which the model should stop. If the population total does not reach the threshold, the Table Select tool returns a value greater than 0, which equates to true. Once this number passes the threshold, it returns a value of 0, which equates to false. Using the Get Count tool gives the model a way to see the number of records. When the Get Count variable is used as the condition variable, the model knows when to stop.

8   Find the Table Select tool and drag it into the model.

9 Use the Connect tool to connect the CountyCensus_Statistics variable to the Table Select tool as the input table. Then double-click the Table Select tool to open the Table Select dialog box. Click the Query Builder button 🖳 next to the Expression entry box. Build the query **SUM_TOTAL_POP < 70000**. Click OK to accept the query and then OK to close the dialog box.

Table Select
Input Table
CountyCensus_Statistics
Output Table
C:\Users\David\Documents\ArcGIS\Default1.gdb\CountyCensus_Stati
Expression (optional)
SUM_TOTAL_POP < 70000
OK   Cancel   Apply   Show Help >>

10 Add the Get Count tool to the model. Connect the output of the Table Select tool to it as the input rows. The new output variable is used as the Precondition variable for the model.

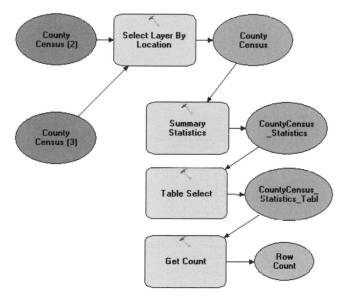

11 Open the model properties and click the Iteration tab. Select "Run the model until this variable is false." Set the variable to Row Count. Enter **10** for the maximum number of iterations. This stops the model after 10 iterations if the variable never returns as false. Click OK to close the model properties.

Census Count Properties

General | Parameters | Environments | Help | Iteration

○ Run the model the following number of times:

1

○ Get the iteration count from this variable:

● Run the model until this variable is false:

Row Count ▼

Maximum number of interations:

10

[OK] [Cancel] [Apply]

12 Save the model. Use the Select Features tool to select one polygon, and then run the model from the model window. After it runs, you may need to click Refresh to see the results.

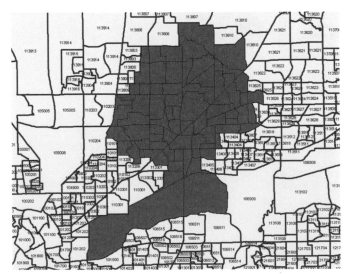

13 Click the List By Source button on the table of contents and find the new summary table. Open it to see the results. If your model did not run successfully, delete the file it created and review the steps to make sure all the necessary components are included in the model and configured correctly.

The table shows the population count, which has exceeded 70,000, and also the total square miles of the selected census tracts. The model can be run many times, using a different starting polygon to generate a large sample of values. Then the city planner will be able to do her calculations and work on her density theory.

This model looks at a cumulative effect from a single starting point. It can be used to summarize flow along lines or networks, or it can be used to look at the effect that the values associated with one group of features has on neighboring features. The results created from using iteration are displayed in your map document so that you can see how an action can have a growing impact as an increasingly larger area is covered.

It is important to note that the Add To Display setting has no effect outside the model window. When you run a model from the Catalog window or the Python window, the Add To Display setting will not be honored. To add model data variables to the display when the model is run from the Catalog window or the Python window, make the data variable a model parameter. Then, on the ArcMap main menu bar, click Geoprocessing > Geoprocessing Options > Add results of geoprocessing operations to the display to enable that option.

What you've learned so far

- How to set up a summary statistic process in a model
- How to control the iterations of a model using a numeric variable and a count process
- How to set a model to respond to a Boolean value
- How to set the maximum number of times to iterate in a model if another method fails to stop it

Iterating in a model by use of feedback

Another type of model iteration allows for the output of a model process to be fed back into the input side of a process. This is known as "feedback." The model is run, and then the resulting feedback is used as the input for the next iteration. Any of the other control routines can be used, whether you run the model a fixed number of times or iterate it until a condition is met. In the model diagram, the feedback path is shown as a dashed line.

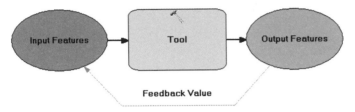

The output of a tool is directed back as input to the same tool in a new iteration.

The variable used for the feedback is set in the properties of the input variable. Note that the feedback can be used only with single-value variables. List or batch variables do not accept feedback. The feedback variable must also be the same data type as the input variable it replaces.

Input Features Properties
General | Data Type | Layer Symbology
This variable contains:
● A single value
○ A list of values
Feedback
Feedback Variable:
Output Features ▾
OK   Cancel   Apply

The feedback variable can only be a single-value output variable.

To control the number of iterations the model will make, an ending condition is set in the model properties. This can be a set number of iterations, a count based on a model variable, or a false condition generated by the model.

This model gets its iteration count from a model variable.

Since the model produces a new output file for each iteration until the ending condition is met, you must create a unique name for each output to prevent the model from overwriting the previous results. This can be accomplished by inserting the system variable %n% into the file name through variable substitution. This variable automatically keeps track of which iteration the model is on. Remember that this variable initializes at 0, so the first output is FileName0.

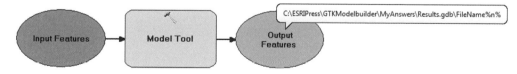

The system variable %n% is inserted to keep the output file names unique.

In the following exercise, as part of the population project the city planner envisions she has also asked you to write a model that creates multiple buffers from an input distance, but that buffers the output of the model a total of five times. When the regular ArcGIS Buffer tool is used, the original selected feature is buffered over and over. If the desired output is several 50-foot buffers, you would enter distances of 50, 100, 150, 200, and so on. By using the feedback option in the model, you can prompt for a single distance and use it to buffer the results over and over—or for this project, five times.

The elements required for this model are simple. In the exercise, you'll need the Buffer tool with its distance value made into a variable and marked as a model parameter. The output variable of the buffer tool will be used as the feedback value for the input variable. Then, you'll need to add the system variable %n% to the name of the output file so that the model saves all the outputs as unique files.

Before you begin the exercise, examine the steps needed to complete the task:

- Create a new model and add the necessary tools.
- Expose a tool parameter as a variable and make it a model parameter.
- Set the model feedback parameters.
- Use the system variable %n%.
- Set the model iteration parameters.
- Save and run the model.
- Examine the results in the Results window and the map display.
- Test the model using different input values.

Exercise 6c

1   Start ArcMap if necessary and open EX06C.mxd. Create a new toolbox called **Chapter 6c** in your MyAnswers toolbox.

2   Create a new model called **BufferTheBuffer** with a label of **Buffer the buffer**. Type a description of **Use the feedback variable to buffer the input feature multiple times.** Be sure the absolute paths check box is selected.

3   Find the Buffer tool and drag it into the model. Make the distance input a variable and mark it as a model parameter. Double-click the Distance variable and give it a default value of **50** feet. Click OK.

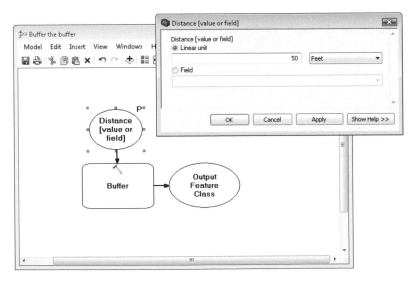

4 Now drag the County Census layer into the model and connect it to the Buffer tool as the input features. **Tip:** Drag the layer on top of the Buffer tool, and then assign its role in the resulting pop-up box.

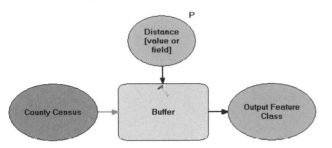

The model is in the Ready to Run state. If you were to run it, it would make a single user-defined buffer around any selected features. The goal, however, is to make multiple buffers, each extending 50 feet from the one before it.

5 Open the properties of the County Census variable and click the General tab. Use the pull-down arrow and set the feedback variable to Output Feature Class. Click OK.

6 In the model diagram, you can use the Select tool to drag the feedback connector line to a more visible location. This adds additional vertices as necessary.

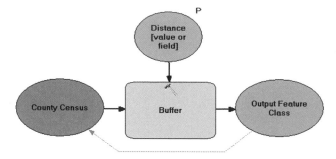

7 Now double-click the Output Feature Class variable to open its input box. Add **%n%** to the end of the default file name. Click OK.

8 Next, right-click the CountyCensus_Buffer%n% variable, and then click Add To Display.

Setting a feedback variable does not by itself cue the model to iterate. As it stands now, the model would run only once. To have it run multiple times, you'd have to set the iteration value in the model properties.

9 Open the model properties and click the Iteration tab. Make sure "Run the model the following number of times" is selected and set the value to **5**. Click OK, and then save the model.

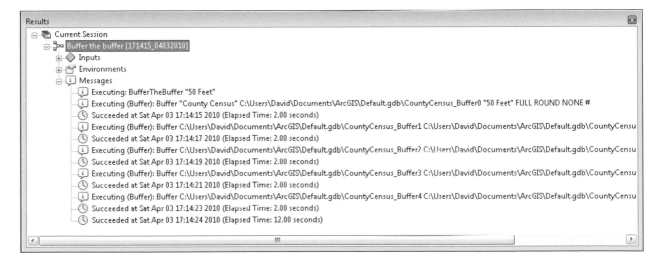

10 Use the Select Features tool in ArcMap to select a census polygon, and then run the model from the model window using the default buffer size of 50 feet. The Results window shows the various iterations. Notice that the input file name changes each time to the output file name with the iteration number appended to the end.

11 To see the results, reverse the order of the output files in the table of contents and zoom in on the selected feature.

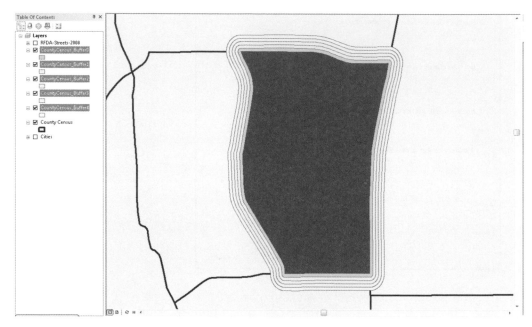

When the model is run from the model window, it uses the default value of 50 feet. It also obeys the Add To Display setting of the output feature class. When the model is run from the Catalog tree, it prompts the user for the buffer distance, but it does not automatically add the output to the table of contents.

12 Save and close the model. Select a different feature in the County Census layer and run the model by double-clicking it. Enter a distance of **140** feet and click OK.

13 Add the results layers back to the table of contents. Zoom in or out as necessary.

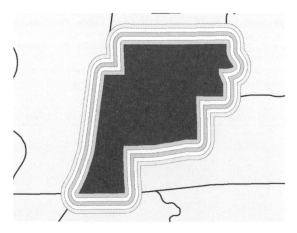

The model has drawn the 140-foot buffers as expected. The city planner will be able to use this model to further enhance her theories on urban sprawl.

Model feedback is an important concept in examining the results of a process and how it affects future iterations of a model. Flow along a network or path may be changed depending on the value found on the current path. Or the selection of one set of features may be used to determine what types of other features are selected in successive iterations.

5

6

7

What you've learned so far

- How to use the results of one run of a model to affect the input values for subsequent iterations of the model
- How to work with the system iteration variable %n%

Iterating in a list

In the previous examples, you use iterations that are set in the model parameters. There are also iterators that can be added to the model and configured similarly to geoprocessing tools. Only one iterator can appear in a model at a time, but the model parameter iterations can be used in conjunction with geoprocessing tools to create very complex models.

One of the simplest iterators is the ability to compile a list and run a process for each of the found values. The tool is given a folder or workspace, and it compiles its own list of what the folder contains. Then the resulting list is used as a batch variable in a model.

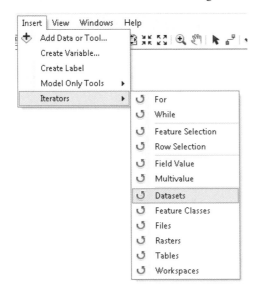

An iterator may retrieve a certain file type, or it may retrieve all the files from a certain location.

These tools can be used to compile lists of datasets, feature classes, files, rasters, tables, or workspaces. For instance, you may want to take all the raster files from a workspace and add them to an existing raster mosaic. The raster iterator makes a list of all the rasters in the input workspace, and then the model repeats the process for each value on the list until all the files are processed.

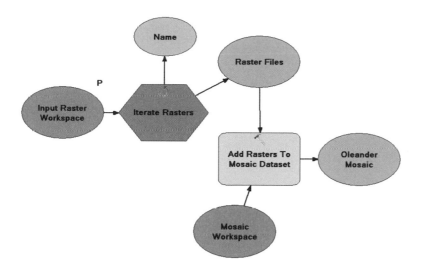

This iterator retrieves raster datasets.

List iterations differ from a standard batch variable in that every file to be processed must exist in the same folder or workspace. But not all the files need to be handled by the iterator. You can set file types, specify keywords in a file name, and in some instances, even specify the feature type. A Recursive option allows the tool to search a specified location as well as all subfolders. In the accompanying example of iteration, only the linear shapefiles from the water data are processed.

Controls can be established to limit the files used as input for the model.

In the following exercise, the member cities of the Regional Fire Department Association have submitted all their new calls for service in the shapefile format. The RFDA board has asked you to perform some spatial statistics and hot spot analysis of the association's entire response area. The tools to perform these analyses work only on feature classes, so in the exercise, you'll need to convert all the shapefiles to feature classes and place them in an existing feature dataset.

Before you begin the exercise, examine the steps needed to complete the task:

- Create a new model and add the necessary tools.
- Add an iterator for feature classes.
- Add the appropriate conversion tool.
- Configure the model to place the results in a feature dataset.
- Save and run the model.
- Examine the results.

Exercise 6d

1   Start ArcMap, if necessary, and open EXO6D.mxd. Create a new toolbox called **Chapter 6d** in your MyAnswers folder.

2   Copy the RFDA Data geodatabase from the Data folder to your MyAnswers folder.

3   Create a new model in your new toolbox and name the model **RFDAConversion**, with a label of **RFDA Conversion**. Add an appropriate description.

4   On the ModelBuilder menu bar, click Insert > Iterators > Feature Classes.

Insert View Windows Help

✛ Add Data or Tool...

Create Variable...

Create Label

Model Only Tools ▸

Iterators ▸ ↻ For

↻ While

↻ Feature Selection

↻ Row Selection

↻ Field Value

↻ Multivalue

↻ Datasets

↻ Feature Classes

↻ Files

↻ Rasters

↻ Tables

↻ Workspaces

5 Open the iterator parameters and set the workspace to the \Data\RFDA Shapefiles folder. Also, set the Wildcard value to *.shp to ensure that only shapefiles are selected. Click OK.

Iterate Feature Classes

Workspace or Feature Dataset
C:\ESRIPress\GTKModelbuilder\Data\RFDA Shapefiles ▾ 📁
Wildcard (optional)
*.shp
Feature Type (optional)
▾

☐ Recursive (optional)

| OK | Cancel | Apply | Show Help >> |

The first part of the model should be in the Ready to Run state.

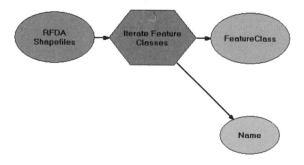

Exercise **6d** Iterating in a list

6 Add the Feature Class to Geodatabase tool to the model. Use the Connect tool to connect the output of the iterator as input features.

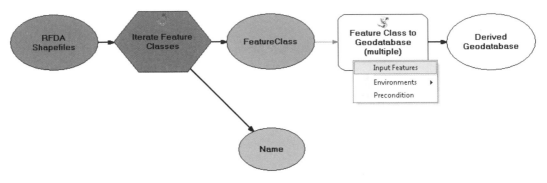

7 Open the parameters of the Feature Class to Geodatabase tool and set the output geodatabase to RFDA Data.mdb\Calls_For_Service_2010 in your MyAnswers folder.

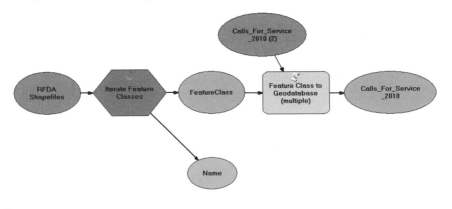

8 Rearrange the tools as necessary to make all the names visible. Right-click the Calls_ For_Service_2010 output data, and then click Add To Display. Then save the model.

9 Run the model from the model window. When the model has finished running, you may need to refresh the map to see the results in the map window. When you have observed the results, close the model window and exit ArcMap.

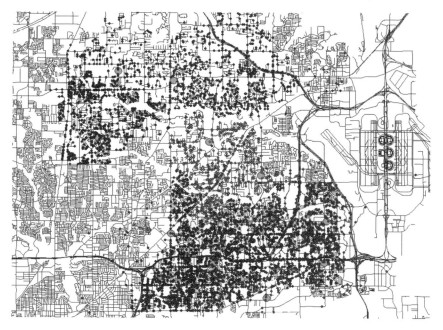

This type of iteration makes the file conversions very easy, whether there are 1 or 100 files. It is important to note, however, that if a model with an iterator is exported to a Python script, the iteration process is not duplicated in the code.

What you've learned so far

- ◆ How to iterate against a list of layers
- ◆ How to work with layer names and types in an iterator
- ◆ How to handle batch output files

Iterating against selected features

Another important type of iteration deals with features or tables in a map document. It involves two types of iterators that go through a selected set of features, one at a time, performing operations on the data. The Feature Selection iterator acts on the selected features in a feature class or feature layer, while the Row Selection iterator acts on the selected records in a table.

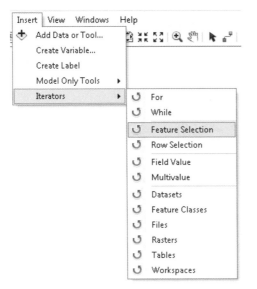

The feature and row iterators move through datasets one item at a time.

The input feature class or the table may have records already selected, and the iterator will work on only the selected features. Any processes linked to the output of the iterator are acted upon as though a single feature is selected.

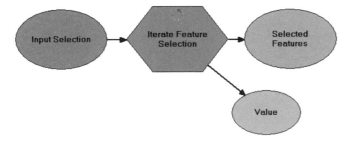

As with most geoprocessing tools, the feature and row iterators operate on only the selected items.

The one-by-one iteration can be time consuming, but it is very thorough. One way to make this type of iteration more efficient is to have the model skip null values. The user can specify what constitutes a null value in the dataset the iterator receives. In the accompanying example, the iterator skips addresses with a value of 0.

Defining what constitutes a null value prevents delays or errors in a model.

This iterator can be combined with the Select Layer By Location tool or the Select Layer By Attribute tool or with a custom Python code block in the Calculate Value tool. Such combinations give this tool greater flexibility for iterating in a feature class or table.

In the following exercise, the library staff at Oleander keeps a log of transactions for all patrons. One of the interesting categories of data they keep is the total number of books each patron has checked out in a given year as well as a running total of all the books a patron has ever checked out. These numbers can be used to determine what percentage of a patron's activity took place in a specified year.

One of the librarians wants to host a contest for children to see who had the highest percentage of checkouts this year. There is a field A that has a code for the type of account, with a value of 2 for children. There is a field C, which is the total number of transactions the patron has conducted this year, and a field D, which is the patron's lifetime total. There is also a field B, but it's not relevant to this project.

In the exercise, you'll need to add a field to the feature class to accept the percentage calculation, and then use field A to retrieve only the children's accounts. Then, you can calculate their transaction percentage this year by using the other fields.

Before you begin the exercise, examine the steps needed to complete the task:

- Create a model to add a new field to the feature class.
- Create a model to perform the feature iteration.
- Set the first model to call the second model.
- Check to see whether the account is a child's (A = 2).
- Calculate what percentage of their lifetime book borrowing occurred this year.

The action of adding a field needs to take place only once, but the verification of the children's code and the necessary calculations need to take place once for each feature. That means these functions need to be in different processes. An easy way to accomplish this is to have one model add the new field, and then call the other model to handle the iterations and calculations.

In the exercise, you will create and test the first model to add a field, and then create and test the iteration model. Finally, you will set the first model to call the second model and pass it the name of the dataset you are working with.

Exercise 6e

1   Start ArcMap and open EX06E.mxd, which is a blank map document. In the Catalog window, copy the OleanderContest geodatabase from the Data folder to your MyAnswers folder.

2   Add the patron activity data for all three years to the table of contents. **Tip:** there is also a dataset called PatronActivity2009Test with a reduced number of features that you should add and use for testing.

3   Create a new toolbox called **Chapter 6e** in your MyAnswers folder. In this new toolbox, create a model called **LibAddField**, with a label of **Library Add Field** and an appropriate description.

4 Find the Add Field tool and add it to the model. Open the tool's parameters and set the input table to PatronActivity2008, the field name to **PercentCheckedOut**, and the field type to Float. Click OK.

5   This part of the model is now in the Ready to Run state. Save and close the model.

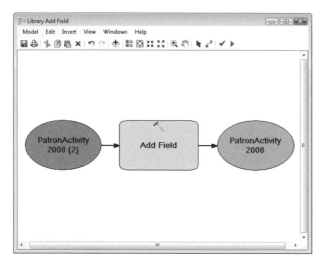

6   In the Chapter 6e toolbox, create a new model with a name of **CalculatePercentage**, a label of **Calculate Percentage**, and a description of **Calculates the annual percentage of activity for youth patrons compared to their lifetime total.** Select the check box to save relative path names.

7   From the Insert menu, add the Feature Selection iterator.

8   Open the iterator's parameters and set the input features to PatronActivity2008. Click OK.

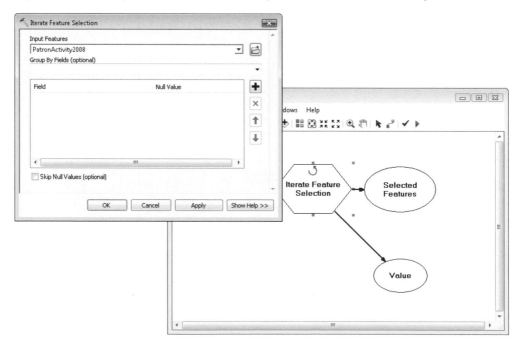

This configuration will now perform iterations for each feature in the feature class.

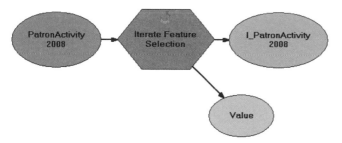

Next, you will need to get the value of field A, and then use the value to determine whether the feature has a code of 2, representing youth accounts.

9 On the ModelBuilder menu bar, click Insert > Model Only Tools > Get Field Value. Open the properties and set the input table to I_PatronActivity2008. In the field input, set the value to A. Change the data type to Long. Click OK.

Get Field Value	
Input Table	
I_PatronActivity2008	
Field	
A	
Data Type (optional)	
Long	
Null Value (optional)	
0	

OK   Cancel   Apply   Show Help >>

10 Change the name of the Get Field Value output variable to **ValueA** so that it is easier to use in variable substitution.

11 Next, insert the Model Only tool Calculate Value.

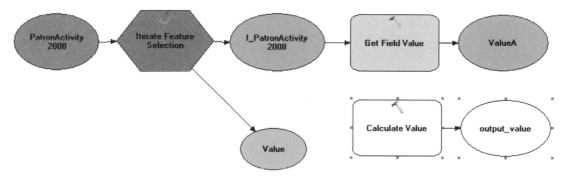

12 Open the parameters of the Calculate Value tool and set the script as shown in the accompanying graphic. The script checks to see whether field A equals 2, the code for children's accounts. Make sure that your capitalization and indentations are correct. Then set the data type to Boolean. This ensures that the output is in a format that the precondition can handle. Click OK to close the Calculate Value dialog box.

One thing to note about using the Calculate Value tool is how it gets input from another tool. You must define a Get statement in the code block section of the tool, and then use the expression to populate the variable. Note that the command getVal(%ValueA%) receives the input in its preset data type. Because you defined this variable as a long integer, the Calculate Value tool will accept it as a long integer. This is an important distinction, because you will be doing a mathematical test on the variable. Otherwise, you could force the input to be a text string by adding quotation marks to the command getVal("%ValueA%").

It is also important to note how the output is used as a precondition. Any precondition looks to see whether the input is true (1) or false (0)—a Boolean value. If you use a number as a precondition, 0 is false, and any other number evaluates to true. In this case, you're sending the actual values of "True" and "False," but you must first define the value as being Boolean before using it as a precondition.

13 To make sure that the field check does not take place before the value is retrieved, make the ValueA output variable a precondition of the Calculate Value tool.

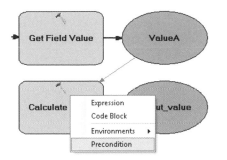

5

6

7

Next, you'll add the tool that performs the actual percentage calculation. It needs to have the output of the Calculate Value tool set as a precondition so that it runs only if the code equals 2.

14 Find the Calculate Field tool and add it to the model. Use the Connect tool to set the input table to I_PatronActivity2008.

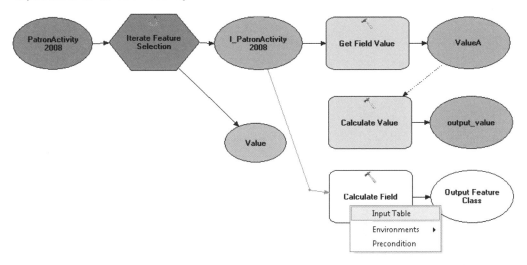

15 Double-click the Calculate Field tool and enter a field name of **PercentCheckedOut**. Remember that this field does not exist until the primary model runs.

16 Click the Expression Builder button ▦. Check that the parser type is VB Script (the default) and build the expression **[C] / [D] * 100**. Click OK to close the Expression Builder. Click OK again to close the Field Calculator dialog box.

Field Calculator

Parser
◉ VB Script   ○ Python

Fields:

[OBJECTID]
[Status]
[Score]
[Side]
[Stan_addr]
[A]
[ADDRESS]
[CITY]
[ZIP]

Type:
◉ Number
○ String
○ Date

Functions:

Abs ( )
Atn ( )
Cos ( )
Exp ( )
Fix ( )
Int ( )
Log ( )
Sin ( )
Sqr ( )
Tan ( )

☐ Show Codeblock

[*] [/] [&] [+] [-] [=]

PercentCheckedOut =

[C] / [D] * 100

[Clear] [Load...] [Save...] [Help]

[OK] [Cancel]

17 Make the output_value variable (from the Calculate Value tool) a precondition of the Calculate Field tool. Only when this variable is true will the percentage be calculated.

18 Finally, make the PatronActivity2008 data variable a model parameter. This is necessary so that the primary model can pass a value to this model. Save and close the model.

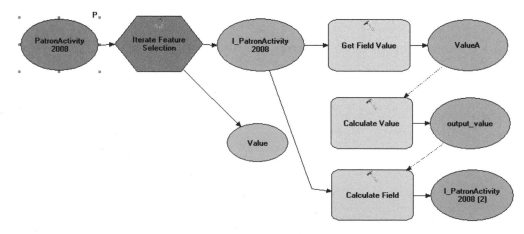

This model handles the iterations and calculations, but you still need to modify the Library Add Field model to accept the name of the feature class from the user so it calls this model.

19 Start editing the Library Add Field model. Drag the model Calculate Percentage from the Catalog window onto the model canvas. Note that it comes with an input data variable already established.

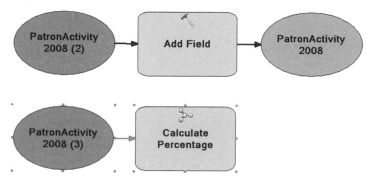

20 Select the variable PatronActivity2008 (3) and delete it. Then use the Connect tool to make the output variable from the Add Field tool the input features for the Calculate Percentage model.

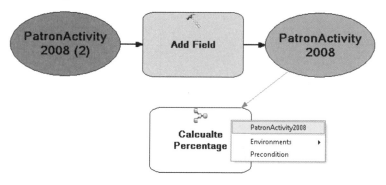

21 Finally, make PatronActivity 2008 (2) a model parameter. Save and close the model. All components are now in the Ready to Run state.

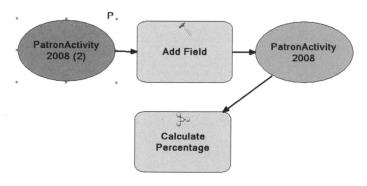

The model runs in the background by default, so that the Results window is not displaying the processes as they run. To get a better feel for how the model is running, you'll want to run it in the foreground.

22 On the ArcMap main menu bar, click Geoprocessing > Options. Clear the Enable check box in the Background Processing area and close the Geoprocessing Options dialog box.

23 Double-click the Library Add Field model to run it. Set the input feature class as PatronActivity2009Test and click OK. The model will take some time to run. If it finishes its run successfully, you can run the larger datasets later.

24 While the model is running, clear the "Close this dialog when completed successfully" check box. When the model has completed its run, examine the geoprocessing Results window. Close the window when you have finished.

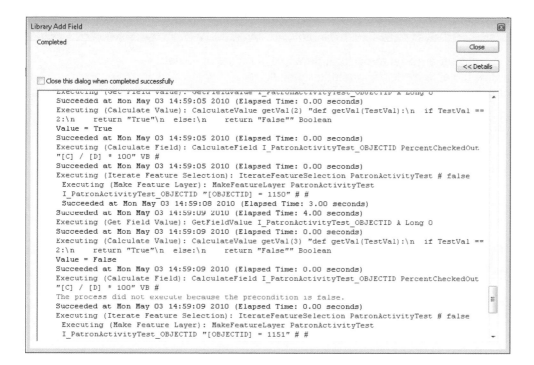

You will notice that the model runs once for each feature in the feature class. When the returned value is true, the model will calculate the percentage. But when the returned value is false (A does not equal 2), the model will generate a message reporting that the precondition stopped the next process from executing.

25 Open the Attribute table for the PatronActivity2009Test layer. Notice that the percentage calculation has been done only for the features with a code of 2 (children).

A	DATE_	B	C	D	E	PercentCheckedOut
1	2003-11-18 19:55:00.000	24720620032826	46	46	2004-09-07 17:03:10.000	<Null>
2	2003-05-09 11:18:00.000	24720610161007	10	11	2004-06-11 16:50:15.000	91.596642
1	2003-11-01 12:51:00.000	24720620035225	1	1	2003-11-01 12:53:06.000	<Null>
1	2003-05-09 11:16:00.000	24720610160991	65	17	2004-07-18 14:39:28.000	<Null>
2	2003-06-21 10:57:00.000	24720620003181	25	45	2004-05-03 10:30:15.000	55.555557
2	2003-04-03 17:39:00.000	24720610122256	11	12	2003-11-30 13:39:59.000	91.666664
2	2003-01-11 12:14:00.000	24720610160090	1	3	2003-10-14 15:42:21.000	33.333332
2	2003-10-16 14:00:00.000	24720620080728	23	23	2004-08-30 19:09:59.000	100

(6 out of 139 Selected)

26 When you are confident that the model is running correctly, run it for each of the other three feature classes to get totals for each year.

What you've learned so far

◆ How to set a model to perform iterations for a set of values
◆ How to call a model from another model

Building structured looping

In order for the ArcGIS ModelBuilder application to function more like a programming language, it includes two looping tools that repeat a process based on values within the model. The first one is the While tool. The While tool accepts a numeric input value and interprets it as either true or false—where 0 evaluates to false and everything else evaluates to true. The user can define which Boolean condition stops the loop.

With numeric variables, a 0 value evaluates to false, and all other values evaluate to true.

In the accompanying example, a buffer is created and used to select property in Oleander. Then a query is done to select only the property with an industrial use and the number of selected features is counted. If the number is 0, the model stops and the next process will not run. If any other number is discovered, the model continues to run.

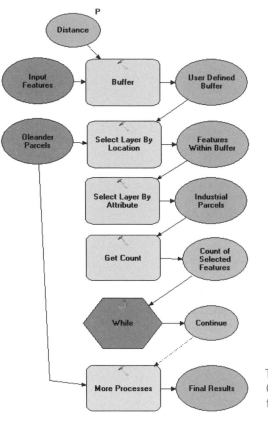

The While statement continues to loop until the Get Count tool returns a 0, which evaluates to false.

This tool seems similar to using the Stop command and the Python script you wrote earlier to check a condition, but it differs in two important ways. If the While command stops the process, it returns to the beginning and allows the user to run the model again. But unlike the Python script that checks a value, it can only check numeric values and evaluate to either true or false.

The other, more complex type of looping is done with the For tool. With this tool, the user defines a starting value (From Value) and an ending value (To Value) along with an increment (By Value) at which the values are changed. The model's process is repeated once for each output value. Note that there is a finite number of times the model can run, and it is by the same increment each time.

For

From Value

1

To Value

100

By Value

10

| OK | Cancel | Apply | Show Help >> |

5

6

7

The For tool uses start, stop, and increment values to loop processes.

The series of output values is fed into the model until the process runs once for each value. In this example, the For statement starts at 2 miles and ends at 10 miles, by increments of 2 miles, to create a series of buffers.

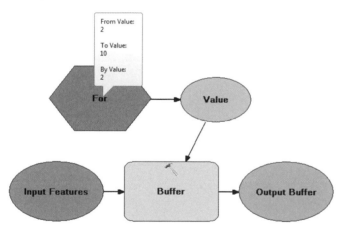

Can you determine how many times this process will loop?

In either of these looping commands, it is important to include a variable substitution on the output feature class name so that the data from each looped process is saved. This can be the system iteration variable %i%, or in the case of the For tool, the value used in the processes, %Value%.

In the following exercise, the library staff is doing some visual analysis of the apartment complexes in Oleander. They want to be able to select an apartment complex, and then create 200-foot, 600-foot, and 1,000-foot buffers around it. In the exercise, you'll create these buffers and overlay them on patron locations so that the library staff can visually judge the apparent walking distances to the library.

Before you begin the exercise, examine the steps needed to complete the task:

- Create a new model.
- Add the For iterator to the model.
- Add the Buffer tool to the model.
- Set the output value of the For tool as the buffer distances.
- Use variable substitution in the output file name.

The user needs to select a parcel before running the model and should expect three buffers every time the model is run. This is characteristic of the For tool, which produces a finite number of results. The user should be able to determine from the model you build that the start distance is 200 feet, the end distance is 1,000 feet, and the increment value is 400 feet.

Exercise 6f

1   Start ArcMap if necessary and open EXO6F.mxd. You will see the Oleander streets, apartment locations, and patron locations from 2004.

2   Create a new toolbox called **Chapter 6f** in your MyAnswers folder to store your new model. Then create a new model in this toolbox named **WalkingBuff**, with a label of **Walking Distance Buffers** and a description of **Generates 200-, 600-, and 1000-foot buffers around the selected apartment complex.**

3   From the Insert menu, add the For iterator, and from the Search window, find and add the Buffer tool.

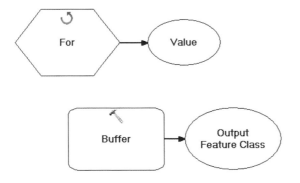

4   Open the For tool and set the parameters for the values discussed earlier.

5 Now open the Buffer tool. Set the input features to ApartmentComplexes and the output feature class to **WalkingBuffer%Value%** in the MyAnswers\OleanderContest geodatabase. Verify that the distance unit is set to feet. Then close the Buffer dialog box and return to the model window.

Notice that the distance value is not set. Therefore, the model is in the Not Ready to Run state. The variable substitution of %Value% also needed to be added to the end of the output file name so that each iteration creates a unique file. It is important to note that the distance unit needs to be set before the Value variable is connected to the Buffer tool. This model uses feet, but you may want to create one that uses miles or kilometers.

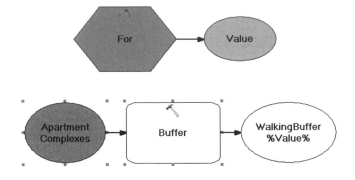

6 Use the Connect tool to connect the Value variable to the Buffer tool, and set it as the distance parameter. The model moves to the Ready to Run state. Right-click the final output variable, and then click Add To Display.

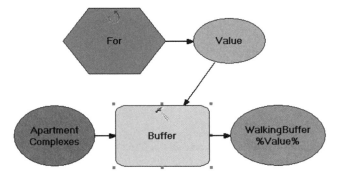

7 Save the model, and then click the Run button on the ModelBuilder toolbar to run it. The model runs three times (as you specified) and creates three new buffers. Close the model window, and then zoom in on the map display to see the results. Note that the distance value was appended to the end of each file name in the table of contents.

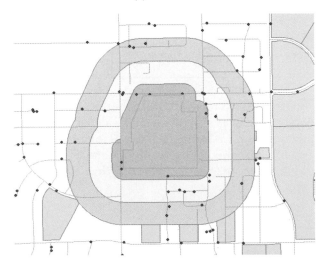

These results let the librarians visually analyze how many patrons live within a certain distance of the apartment complex. Thus, any activities planned at the complex can benefit the apartment dwellers as well as the identified patrons who live nearby.

What you've learned so far

- How to define the values for the For tool
- How to integrate variable substitution into an iterated process

5

6

7

The iterators in this chapter illustrate a wide array of techniques for using models to process data. Remember that only one iterator can be used per model, but models can be called inside other models, letting you nest an iterator inside another iterator. Just be careful to use variable substitution with the output file names to keep these names unique.

Chapter 7

Building model documentation

An important aspect of modeling is the need to document the model elements and provide information to the user on the expected inputs and outputs for a model. Using the model documentation tools, a programmer can provide users with this information as well as record programming notes for future reference.

Building in-model documentation

If a model that has been in use for some time requires updating to accommodate a new software release, new tools, or changes in its purpose, it can be difficult for the programmer to remember the design process behind all the elements. This is especially true if a different programmer has the task of updating an existing model. Moreover, including an attractive graphic of the model diagram is a useful addition to the user Help.

All the elements in a model display text inside their symbol representation. Making these element names readable and descriptive of the element's purpose adds greater clarity to a model. One way to accomplish this is to rename the element by going to the element's context menu and clicking Rename.

A descriptive name makes it easier to keep track of variables.

Any element can be renamed, but note that if a stand-alone variable is renamed, it is the new name that will be used for in-line variable substitution. If the substitution has been done before the renaming, the user will need to update the tool that is using the variable before running the model.

Variables may also need to be resized so their entire name is displayed. If the name is larger than what fits in the display, the symbol can be enlarged. When an element is selected, a set of "resizing handles" is displayed with blue boxes in all the corners. Dragging one of these boxes resizes the element.

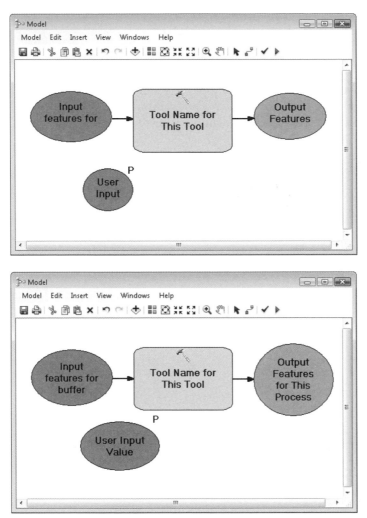

Elements are resized so their entire label is visible.

The elements may also need to be moved around the model canvas for clarity. When an element is moved, any connector lines will be redrawn to maintain the connection. By clicking these lines, a new vertex may be added and the lines can be adjusted for better visibility.

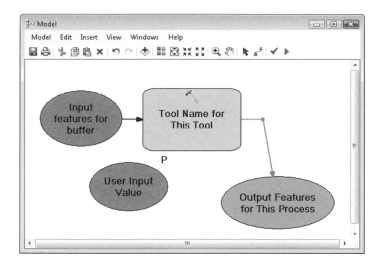

Connector lines show the order of processing in a model.

Another way to add documentation to a diagram is through the use of labels. Labels can be either freestanding or associated with a particular element. Stand-alone labels can be placed anywhere, but they may need to be adjusted either manually or with the Auto Layout tool if any elements are moved. Labels associated with an element move with the element, requiring less cleanup after a model is modified.

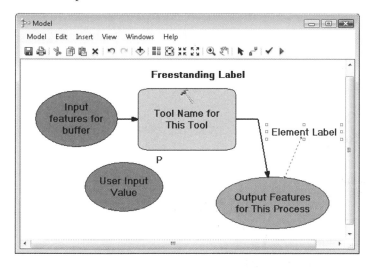

Labels can be added outside an element to add more information.

Exercise **7a** Building in-model documentation

The properties of the labels can be altered to give them different looks within the diagram. Some may be used as titles and some as general notes or comments, depending on their size or color. The Display Properties dialog box is used to make these changes.

Display Properties	
Selected Label Properties	
Name	Element Label
Tooltip	
URL	
Font	MS Shell Dlg
Text Justification	Center
Resizability	Tight Fit
Background Color	
Border Color	
Border Width	
Transparent	True
Width	99
Height	16
X Center	450
Y Center	238
Orientation	Outside
Region	Any
X Absolute	0
Y Absolute	60

A label's font, color, alignment, and more are controlled by the display properties.

By controlling the characteristics and placement of the labels, the diagram can be made to contain information about the processes as well as other general information about user interaction with the model. Colored boxes can also be added via the display properties to add emphasis.

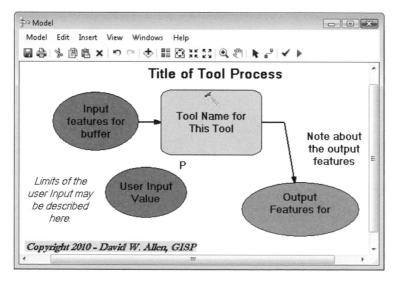

This model is well documented through the use of different label styles.

When the model diagram is exported as a graphic, all the labels and customized settings are visible. This can make for an attractive and informative diagram, which can be used in later documentation.

In the following exercise, you have completed the Select By Code model and it's in need of some documentation. In the exercise, you'll start by cleaning up the model diagram and adding labels to the model elements. Then you'll add a title and some general notes describing the model's uses.

Before you begin the exercise, examine the steps needed to complete the task:

- Start editing the model.
- Resize a variable component.
- Create a stand-alone label.
- Change the label's display properties.
- Create a label tied to an element.
- Create additional stand-alone labels and change their display properties.
- Create a graphic of the model diagram.

Exercise 7a

Adding labels and arranging the model elements will make this model more visually appealing and easier to understand. These additions will also be visible in the model graphic that you create at the end of the exercise.

1 Start ArcMap, if necessary, and open EX07A.mxd. Copy the Chapter 7a toolbox from the SampleModels folder to your MyAnswers folder.

2 Start editing the Select By Code model. Right-click the OleanderCalls (2) variable, and then click Rename. Change the name to a more manageable one, **Oleander Calls for Service**. Click OK.

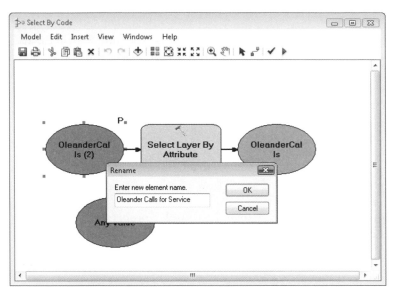

3 Select the Any Value variable, rename it **CodeNum**, and resize it so that the entire name appears on a single line. Remember that renaming this variable affects any variable substitution that uses it. It will now be referenced as %CodeNum%.

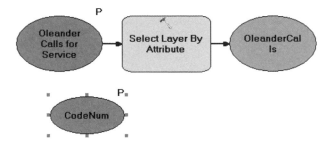

4 Select and resize the OleanderCalls variable so that the name appears on two lines. Note that even though there is not a space in the name, it can be made to display correctly on two lines.

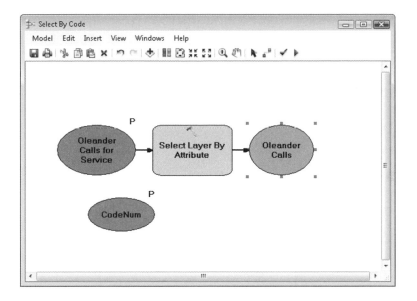

5 On the ModelBuilder menu bar, click Insert > Create Label. This adds a new stand-alone label, which will become the model's title.

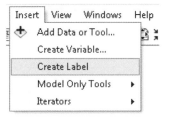

6 Right-click the new label, and then click Display Properties.

7 On the first line, click in the Name row and enter **Select by Incident Code**.

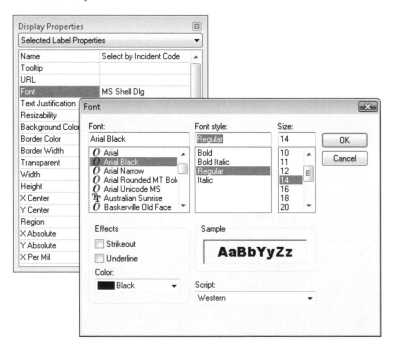

8 Next, click in the Font row, and then click the Ellipsis button [...] to open the Font dialog box. Set the font style to **Arial Black, Regular, size 14**. Click OK, but leave the Display Properties dialog box open. Move the label into position at the top of the model as necessary.

Take a minute to examine the other label properties you could change. These include outline colors, ToolTips, hyperlinks, and text justification, to name a few.

9 Right-click the CodeNum variable, and then click Create Label. Since this action originated with the element itself, the label is tied to that element.

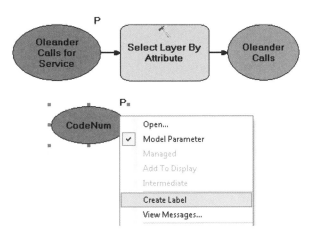

10 The label is created over the element. Click in an open area to clear the selection, and then click the label and drag it just below the variable. Note that as you drag the label, a dashed line shows which element it is associated with.

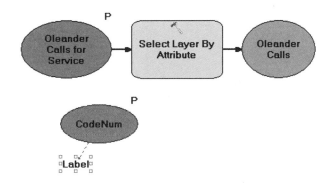

11 The Display Properties dialog box is updated to the selected feature automatically. Change the name to **Incident Code** and press ENTER. Note that if you move the CodeNum variable, the label moves with it.

Select by Incident Code

12 Now add another stand-alone label. Make it a general note. Using the Display Properties dialog box, change the font to **Arial, Italic, size 10**. Then click Border Width and select the thinnest line as the border style.

Select by Incident Code

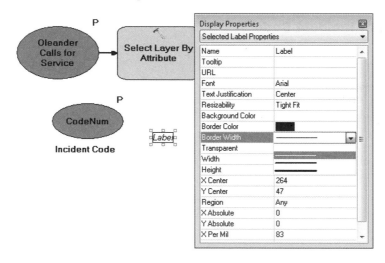

13 Although the name could be changed in the Display Properties dialog box, the dialog box does not allow for multiple-line text. Double-click the label and type **The user will input**. To go to the next line, hold the Shift key and press ENTER. Then type **an incident number,**. Add another line and type **which will be selected**, and on a fourth line type **and displayed.** Press ENTER to finish and move the label into place as necessary.

Select by Incident Code

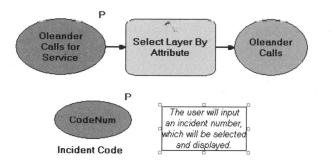

14 Finally, create another stand-alone label, which reads **Copyright 2010 -** and your name. Then in the Display Properties dialog box, click Background Color and select a color of your choice. Click OK, and then close the Display Properties dialog box. This puts a colored box around your copyright notice.

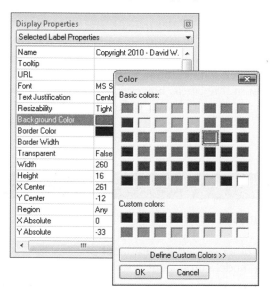

15 Create a graphic of the model called **SelectIncident** and save it in your MyAnswers folder. Then save and close the model.

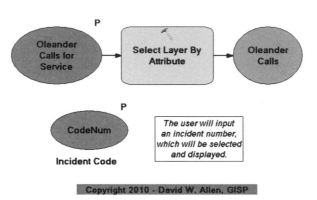

Select by Incident Code

Renaming and resizing model elements makes them easier to read, and their labels help to describe their functions. By adding labels with different fonts and colors, you can annotate the model for future reference. Some labels are freestanding, but others are tied to a model element. The labels tied to a model element move with that element, even when you use the Auto Layout tool. Freestanding labels, however, do not move with the Auto Layout tool, so make sure that your model layout is finished before adding stand-alone labels.

What you've learned so far

- ◆ How to change the shape and size of a model component
- ◆ How to create a stand-alone label
- ◆ How to change the font, size, color, and other properties of a label
- ◆ How to create a label tied to an existing model component

Creating user Help messages

The labels and annotation created inside a model are a great way to document the processes from the programmer's perspective, but the user will not see them. The user runs the model and is presented with a parameters dialog box. This interface gives the programmer an opportunity to display Help messages about the model so that the user will better understand how the model functions and what input they should supply. The most basic of the Help messages is the model description that is added in the model's Properties dialog box.

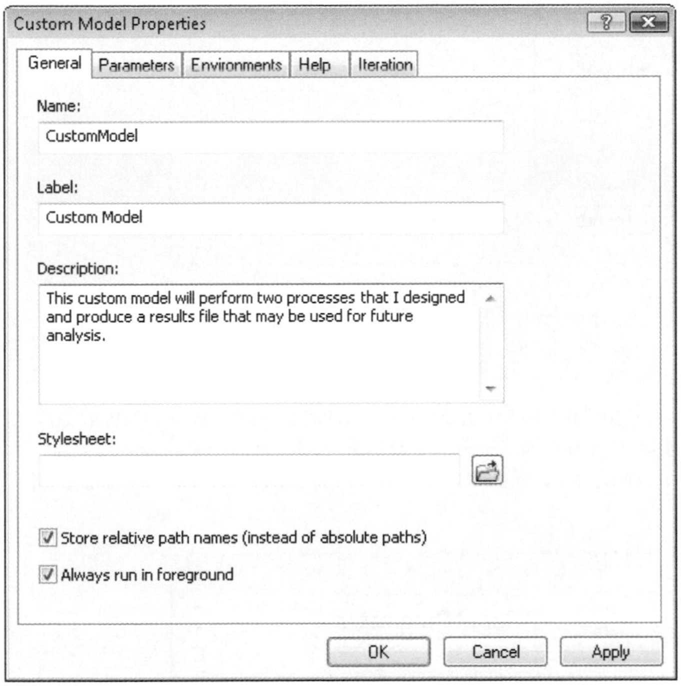

5

6

7

A model's description can provide a basic Help message.

This message is displayed when the model runs and appears when Show Help is clicked. This message is limited to simple text, and the programmer has no control over fonts, sizes, colors, or the general layout.

A single Help message is displayed by default from the model's properties.

There are no additional Help messages provided for the tool's input boxes. While clicking an input box on a standard ArcGIS tool produces a description of the type of data that is required for that input, the model's Properties dialog box does not allow for the addition of this type of Help message.

The default Help message does not include input-specific help.

The Item Description dialog box, however, allows the programmer to build a rich Help display that can include various fonts, colors, bulleted lists, images, and more. This is accessed by right-clicking the model name and then clicking Item Description.

The Item Description dialog box can be used to build a more complex Help display.

The categories that can be changed in the item description are title, summary, usage, syntax, code samples, tags, and credits. The title area lets you give your model a title other than the label listed in model properties. It also accepts an image to help illustrate the model. A graphic of the completed model diagram or a thumbnail depiction of a critical dataset can be included here.

Item Description

An image of the completed model may be helpful in understanding the model's overall objective.

The Tags area lets you type keywords about your model and the processes it performs. These can be one- or two-word terms, separated by commas, to provide a quick summary of the tool that can be indexed.

Tags

Select sewer lines, buffer by pipe size, sewer utility, City of Oleander, Public Works, summarize length

The user can define tags to be used in the ArcGIS tool index.

These keywords are indexed in the Catalog window so that in any searches you perform you will find your model as a valid tool. This makes your custom model discoverable in the Search window just like the standard ArcGIS tools.

Search   ⇥ ✕

← →   ⌂ ⟳ ▤ Local Search   ▾

ALL   Maps   Data   **Tools**

sewer   🔍

Search returned 2 items.   Help

Linear Referencin   My Custom   ⊗
Summary: not ava   Model for
toolboxes\system t   Running Multiple
                    Processes
My Custom Mode
Processes   **Type:** Tool
Summary: not ava
toolboxes\my toolk   **Tags:** Select **sewer**
                    lines buffer by pipe
                    size **sewer** utility
                    City of Oleander
                    Public Works
                    summarize length

                    Item Description

This custom tool was found in the index through one of its tags.

Next is the Summary area. This area accepts a long description of exactly what the model can be used for and any preparations that may be needed before the model is run. The Edit pane for the summary allows you to use formatted text. This includes font settings, bulleted lists, indentations, and several alignment options.

Summary

- A valid sewer line database
- A set of street centerlines
- A symbology template layer

Running this model will produce a sewer line summary. Note that all lines in excess of 48" in diameter are not maintained by the City of Oleander and should not be included in the maintenance budget.

The Summary area includes a variety of formatting options.

The Usage area, like the Summary area, provides a set of text formatting tools. This area typically contains information directly related to how the model is used rather than a description of how it works.

Usage

1. If any edits have been made to the data since the last time this model was run, you will need to calculate the length in feet and meters into the LengthFT and LengthM attribute fields.

   Right-click the field and select Calculate Geometry

2. Verify that the latest as-built drawings have been scanned and indexed to the sewer line data. The file names will be included in the output database.

The Usage area can also be formatted and should contain the steps necessary to prepare and run the model.

The model generates the framework of the Syntax area automatically, but it contains only generic information about how the model might be used in a script. You can also add context-sensitive Help to each of the model's parameters by providing an explanatory phrase for each input parameter. This form of Help also allows for the use of formatted text.

Syntax

⌃ Input_Parameter

Dialog Explanation

Please provide the inch-diameter-mile conversion ratio as calculated by the ADSM.

- This number represents the City's discount rate for sewage treatment based on infiltration into the system.
- Check with the Director of Public Works for the most current value.

The symbology of this input file will be converted to cartographic representations in the output file.

Scripting Explanation

The Syntax area can be used to build context-sensitive Help for each required input.

The Syntax area is displayed in the Tool Help as the tool runs. Each input shows a different element of Help when its input box is selected.

Each of these input fields displays a unique Help screen.

Any model or script you create can be used in another custom script, so providing code samples is a good way to help others understand how this script can be integrated into their own scripts. An easy way to do this is to export your model to a script, and then cut and paste the resulting code in the Code pane of the Code Samples area.

Code Samples

```
(^) Title    Sewer system summary                                    ✕

B   I   U   A⁺  A⁻  ⋮≡  ⋮≡  ≡  ≡  ≡  ≡  ⋞≡  ⋞≡  ↻  ↺

This custom model will perform two processes that I designed and produce a results file that may be used for future analysis. ▲

                                                                      ▼
```

Code

```python
# Import arcpy module
import arcpy

# Script arguments
Input_Parameter = arcpy.GetParameterAsText(0)
if Input_Parameter == '#' or not Input_Parameter:
    Input_Parameter = "100" # provide a default value if unspecified

Input_Feature_Class = arcpy.GetParameterAsText(1)
if Input_Feature_Class == '#' or not Input_Feature_Class:
```

The code sample can illustrate how the model's processes are coded in Python or another programming language.

And finally, give yourself a little credit for crafting such a powerful script. The Credits area should include the name of the script author and any restrictions on the script's use. This may include a disclaimer or copyright statement.

Credits

```
Author: David W. Allen, GISP
Date: July 2010

This model is placed in the public domain by the City of Oleander and may be freely distributed. It is not supported, and may not be suitable for your data.
```

The credit information is important if the model is to be shared with others.

It is important to take the time to document your models. This is valuable information for anyone who wants to adapt your model to their own situation. It might even be helpful to you if you are required to do modifications to the model at a later date. If the model is to be run by other city staff members, they will appreciate the context-sensitive Help and other information you provide in your documentation.

The final Help file you produce will look strikingly similar to the standard ArcGIS tools. The reason is that Esri programmers use the same type of dialog to document the system tools.

Custom Model

Title   **Custom Model**

Summary

This will describe how the model works. Before running this model, be sure to have these things ready:

- A valid sewer line database

- A set of street centerlines

- A symbology template layer

   Running this model will produce a sewer line summary. Note that all lines in excess of 48" in diameter are not maintained by the City of Oleander and should not be included in the maintenance budget.

Illustration

Usage

1. If any edits have been made to the data since the last time this model was run, you will need to calculate the length in feet and meters into the LengthFT and LengthM attribute fields.

   Right-click the field and select Calculate Geometry

2. Verify that the latest as-built drawings have been scanned and indexed to the sewer line data. The file names will be included in the output database.

All the different sections are displayed in the custom Help screen.

All the labels and model properties that are set in the model window can be seen only when the model is edited. None of that information helps when the model is being run. In the following exercise, you'll add documentation that populates the Help screens for the tool.

Before you begin the exercise, examine the steps needed to complete the task:

- Change the model's description.
- Edit the Help file.
- Populate the various Help categories.
- Save the documentation with the model.
- Test the results.

Exercise 7b

Models and scripts have a special editing dialog box where you can build the Help files. You'll open this editor and populate it with all the Help information. **Tip:** If you want to avoid retyping all this documentation, the text can be found in the Select By Code documentation in the SampleModels folder. You can copy and paste the information into the Item Description dialog box, but you'll need to pay attention to the formatting.

1   Start ArcMap, if necessary, and open EX07B.mxd. Continue working on the model you worked on in exercise 7a, or copy the Chapter 7b toolbox from the SampleModels folder to your MyAnswers folder.

2   Right-click the Select By Code model, and then click Item Description.

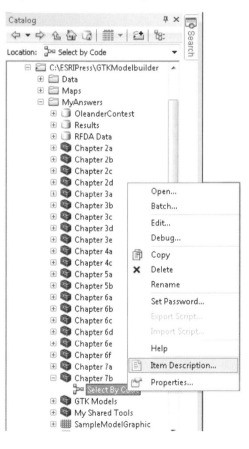

5

6

7

3 This blank template will be filled with information after you have finished. Click the Edit button ![Edit icon] Edit at the top of the Item Description - Select By Code dialog box.

4 In the Title area, type **Select calls for service by incident type.**

Item Description

Title   Select calls for service by incident type.

✖ Delete   ![Update icon] Update...

5 Click Update and browse to the graphic you made of your model in exercise 7a, or use the Select By Code diagram in the SampleModels folder.

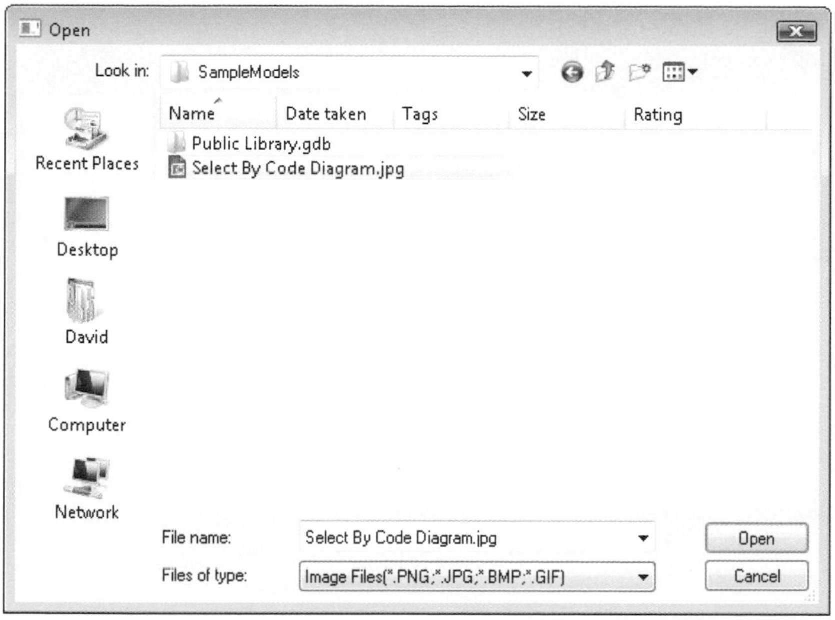

6 Add the terms shown in the accompanying graphic, separated by commas, in the Tags area of the Description window.

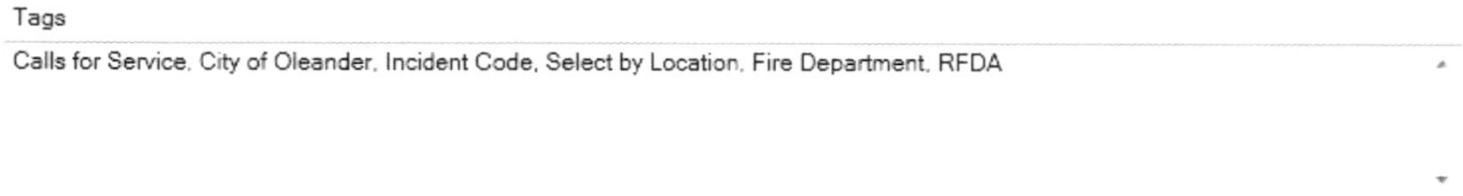

Tags

Calls for Service, City of Oleander, Incident Code, Select by Location, Fire Department, RFDA

7 The Summary area allows for text formatting. Add the first line as bold-faced and then the next three lines in a bulleted list. Finish with the last line in italics.

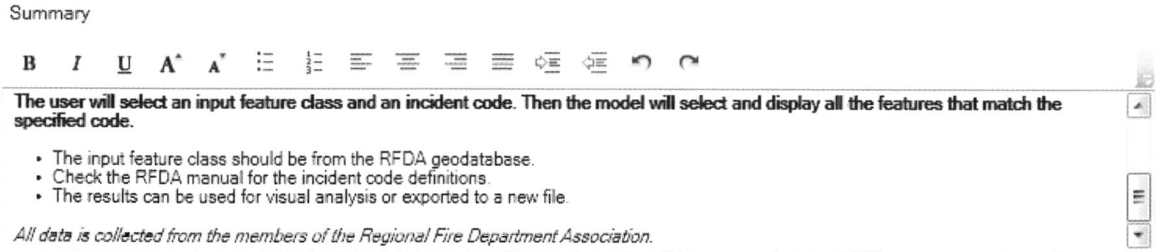

Summary

B *I* <u>U</u> A˄ A˅ ≔ ≟ ≡ ≡ ≡ ≡ ⫣ ⫤ ↶ ↷

The user will select an input feature class and an incident code. Then the model will select and display all the features that match the specified code.

- The input feature class should be from the RFDA geodatabase.
- Check the RFDA manual for the incident code definitions.
- The results can be used for visual analysis or exported to a new file.

All data is collected from the members of the Regional Fire Department Association.

8 In the Usage area, add the information shown in the accompanying graphic. This area also allows for text formatting.

Usage

The model should be run for the following types of analysis:

1. Locate all the features matching one type of incident
2. Identify the regional response areas of the RFDA
3. Locate calls for service outside a city's normal response area

The Syntax area is also known as context-sensitive Help. When the user clicks in the data entry boxes for the model's required parameters, the information you entered in this space will be displayed.

9 The Syntax area shows both input variables, each of which can accept documentation. Expand the input area for CodeNum.

Syntax

⌄ CodeNum

⌄ Oleander_Calls_for_Service

10 Enter the information for the CodeNum attribute. Be careful to match the formatting in the accompanying Syntax graphic.

Syntax

⌃ CodeNum

Dialog Explanation

Enter an incident code to select.

Check the RFDA manual for the incident code definitions.

Contact Chief Isbell at the RFDA office at (555) 685-1628.

Scripting Explanation

The data type for the incident code is numeric.

11 Expand the input area for Oleander_Calls and enter the information as shown in the accompanying Dialog Explanation graphic.

⌃ Oleander_Calls

Dialog Explanation

B *I* U A⁺ A⁻ ☰ ☷ ☰ ☰ ☰ ☰ ⇥☰ ⇤☰ ↺ ↻

Select the feature class containing calls for service features.

The calls for service data from the RFDA members is already formatted for this analysis. If you use data from other agencies, confirm the field names and codes for the incident types.

Scripting Explanation

B *I* U A⁺ A⁻ ☰ ☷ ☰ ☰ ☰ ☰ ⇥☰ ⇤☰ ↺ ↻

12 Click Save and notice the text formatting and graphic in the Description page. Close the page when you have finished.

Description

🖨 Print   ✎ Edit

Select By Code

Title   **Select calls for service by incident type.**

Summary

The user will select an input feature class and an incident code. Then the model will select and display all the features that match the specified code.

- The input feature class should be from the RFDA geodatabase.
- Check the RFDA manual for the incident code definitions.
- The results can be used for visual analysis or exported to a new file.

All data is collected from the members of the Regional Fire Department Association.

Illustration

Usage

The model should be run for the following types of analysis:

1. Locate all the features matching one type of incident
2. Identify the regional response areas of the RFDA
3. Locate calls for service outside a city's normal response area

Syntax

SelectByCode (CodeNum, Oleander_Calls_for_Service)

13 Edit the Select By Code model in the model window and export it to a Python script. Name the file **SelectByCode.py** and save it in your MyAnswers folder. Close the model window.

14 Open the Item Description - Select By Code dialog box again and begin editing the description information. Click Add Code Sample.

15 Type the title and description, and then copy and paste the Python code in the exported script you created in this exercise into the Code area.

16 Finally, type your name as the model's author and add the date and the disclaimer shown in the accompanying Credits graphic.

Credits

Author: David W. Allen, GISP
Date: July 2010

This data was compiled by the Regional Fire Department Association, with headqurters in Euless, Texas.
The data is released into the public domain. Credit 'RFDA of Texas.'

17 Check all your entries in the Item Description - Select By Code dialog box and click Save.

18 To test the model documentation, double-click the model to run it. Click Show Help, if necessary, to expand the Help section.

Select By Code

CodeNum
100

Oleander Calls for Service
OleanderCalls

Select By Code

The user will select an input feature class and an incident code. Then the model will select and display all the features that match the specified code.

- The input feature class should be from the RFDA geodatabase.

- Check the RFDA manual for the incident code definitions.

- The results can be used for visual analysis or exported to a new file.

All data is collected from the members of the Regional Fire Department Association.

OK   Cancel   Environments...   << Hide Help   Tool Help

5
6
7

19 Now click in the CodeNum text box. The Help will change to match the context of your position in the Select By Code dialog box. You can also check out the context-sensitive Help for the Oleander Calls for Service text box.

20 Click Tool Help at the bottom of the Select By Code dialog box. This opens an HTML-based Help page that includes all the information you provided.

21 When you have examined the Help, close the Help screen and cancel the model.

With a little effort, you've made a pretty impressive and comprehensive Help page. There is one other feature you can add to your model that will make it function more like the ArcGIS system tools. This is the ability to locate the model through the Search window.

22 Click the Search tab to open the Search window, and then click the Index/Search Options button 🔳 . This dialog box is used to add your custom models and scripts to the tools index.

23 Click the Index tab. Then click Add and navigate to the Chapter 7b toolbox in your MyAnswers folder. Click Select.

24 Next, click Index New Items. The index is automatically updated to include your new selection. Any of your other toolboxes or folders can be added and indexed, but be warned that the indexing process can take some time.

25 When the index is finished, click OK to close the Index/Search Options dialog box.

Once the index is built, any searches will search the tags you put in your model documentation.

26 In the Search text box, type **RFDA**. Your new model is found and displayed in the Search window.

27 Close ArcMap.

As you can see, adding documentation and a complete, indexed Help file can make your model act just like the ArcGIS system tools. Although this may be the last part of the model you work on, and is the last part of this book, adding documentation is an integral part of making your model complete.

What you've learned so far

- How to create an HTML-based Help page for a model
- How to build a Help display that includes text, illustrations, bulleted lists, and code samples
- How to set up context-sensitive Help for a model's input parameters
- How to include custom tools and scripts in the Search index

This concludes the chapter content and exercises in this book. If you have worked through all the material, you should now have a better understanding of what models are, how they are constructed, and how they can be used to streamline your workflows. You should also have a better understanding of the different elements that are used in models, from variables and model parameters to looping and branching.

At this point, you can move on to appendix A and work through the challenges it provides. These exercises will give you extra practice building models. They come with a defined outline and use the student data that comes with this book. The techniques you learned in the exercises are reinforced in these challenges.

Next, you will want to begin turning your own workflows into models. Start with something simple, such as creating a custom tool that automates a process. Then try adding model parameters to make the process more flexible and able to handle different situations. Try using variables and variable substitution in your model. This involves using the basic components of modeling, and it should be easy to accomplish.

Once you've mastered these skills, move on to more complex concepts such as batch processing and using system variables to track the output file names. Try making your model act on multiple files or have it accept multiple inputs from the user that will then act together on a single file. Practice building models that use iteration, call scripts, or call other models. You could also try incorporating simple error checking and branching with the use of the Stop tool.

The final hill to climb is to start writing Python scripts. Use the book's examples to start with simple error checking, branching, and calculations. Later, after some practice, take on more complex scripts. Remember, though, that much of what you need to do in a script can be done in the ArcGIS ModelBuilder application, with the scripts just serving as a control to check values or evaluate Boolean expressions. And if you need the end product to be a Python script, try building a model to accomplish that task, and then export the model to a script.

As you start building more complex models, you will see how easy it is to automate your workflows and build useful applications. With your ability to create models that can call other models or scripts, iterate over features and datasets, and harness the power of Python scripting, your world of modeling will continue to expand as you find more and more ways to adapt this exceedingly versatile technology to your own field of expertise.

Model challenges

The following five challenges are modeling scenarios for you to work on, on your own. These challenges reinforce the topics of selecting features, branching, using iterators, scripting with Python, using multiple inputs, and incorporating geoprocessing tasks.

For each challenge, an overall description and outline of the steps needed to complete the task are given, along with a description of any data to be used. Read the challenge, and then design the processes and create a model to perform the tasks.

None of these challenges involves real-life scenarios, but they should be good practice for using the modeling tools.

Challenge 1: Calculating tax values

The economic development director for Oleander, Ms. Houston, has a project that deals with the tax revenues from single-family housing. She would like to be able to enter a tax rate, and then apply it to the current land values and see what the total tax revenues would be.

The tax value data is in a feature class called TaxValues in the Property Data feature dataset, which is in the CityOfOleander geodatabase within the Data folder. In the exercise, the most current data hasn't come out yet from the tax office, but you can build and test the model using the field TaxVal06. The tax rate for Oleander is 49 cents per \$100 of valuation. The calculation then is (TaxVal06 × .0049). Ms. Houston will want to enter different tax rates and see the corresponding change in revenues. For this exercise, build in some error checking to make sure that the rate entered is between 25 cents and 75 cents. If the number entered is out of this range—for instance, 0.25 instead of 25—correct the value and continue with the calculation. You can isolate the single-family properties by finding all the features where the field DU (dwelling units) = 1.

Before you take on the challenge, examine the steps needed to complete the task:

- Prompt the user to enter a tax rate—build good documentation for this.
- Perform error checking to make sure that an appropriate value was entered.
- Select all the single-family property (DU = 1).
- Perform the calculation and display the results.

You can also test the model on the other tax value data provided in this feature class.

Challenge 2: Creating a library readership index

This project involves going through the data for the Oleander Library and assigning a readership index so the city can reward the best patrons. It entails writing a model to find patrons by their classification number, and then performing a calculation for each patron to determine the readership index. It uses iterators to go through the datasets and a Python script to control the flow and perform the calculations.

In the exercise, the library has gotten a bookstore chain to sponsor a reading contest and has developed a formula for calculating a readership index. The library wants to do the calculation for 2010 data and determine the winners, who will receive an award from the sponsor. There are awards in the adult and youth categories and for the overall winner based on the highest index number. The calculation is as follows:

- 1 point is awarded for each book read, up to 49 books.
- 2 points are awarded for each book read from 50 to 99 books.
- 3 points are awarded for each book read from 100 to 499 books.
- 4 points are awarded for all books at 500 and above.
- For youth patrons, the points are doubled.

You can find the PatronActivity2010 dataset in the Public Library geodatabase, which is in the SampleModels folder. The field Classification identifies adult patrons with a 1 and youth patrons with a 2. The field Annual Total shows the number of transactions the patron has conducted over the year. The field Patron ID will be used by the librarians to identify the winners from their master patron registration database, which we do not have access to.

Before you take on the challenge, examine the steps needed to complete the task:

- Allow the user to specify the dataset to work with.
- Add a field to contain the results of the calculation (what data type should it be?).
- Set up an iterator to iterate through the features in the feature class.
- Determine whether the current record is classified as adult or youth.
- Perform the points calculation (remember to double it for youth):
  - If the number of transactions is less than 50, the calculation is Annual Total x 1.
  - If the number of transactions is between 50 and 99, the calculation is ((Annual Total - 49) × 2) + 49.
  - If the number of transactions is between 100 and 499, the calculation is ((Annual Total - 99) × 3) + 149.
  - If the number of transactions is at 500 and above, can you determine the calculation?
- As a bonus, make all the point values variables so they can be assigned different values for each run of the model (stand-alone variables used as model parameters inserted into the calculations with variable substitution).

When the model has completed its run correctly, determine the patron ID of the adult and youth winners, as well as who the overall winner is. Save the model for use in future years.

Challenge 3: Developing fire department street guides

Chief Isbell of the Regional Fire Department Association (RFDA) has been looking at fire response times in the area. Generally, he's pleased, but he's identified an issue when firefighters respond to a mutual-aid call from a station other than their home station. In the exercise, he wants you to generate a list of streets within one mile of each station so that firefighters can study the response times. This involves using an iterator to get the station locations and a selection process to get the surrounding streets.

The RFDA already has the data necessary for this operation. You'll need to first start an iterator that selects each station, and then use the Select By Location tool with an optional distance to select the street centerlines. This step returns all the street segments, with many of the street names duplicated in the selected set. To get a list where each street name is listed only once, you'll need to perform a summary statistic operation. The necessary data is in the Base Data feature dataset, which is in the RFDA Data folder within the Data folder. The layer RFDAStations has a point for each station. The file RFDA_Streets_NCTCOG contains the street centerlines, and the field LABEL is the field where you'll do the summary of the street segments to get the street names.

Before you take on the challenge, examine the steps needed to complete the task:

- Use a feature selection iterator to get each station in the RFDA district.
- Use the Select By Location tool to select within a distance of each station and specify 1 mile. As a bonus, you could make the distance a model parameter for future analysis.
- Use the Summary Statistics tool to output a new table for each station. **Tip:** Be sure to include a system count variable such as %n% in the output file so that each name is unique. As a bonus, use the Get Field Value Model Only tool to retrieve the value of the field Title for each station and include it in the name of the output file.

When the model is running correctly, you will be able to produce a street list for each station and control the search distance for generating the list of names.

Challenge 4: Working with tax value statistics

The economic development director for Oleander, Ms. Houston, would like to do a little value analysis of residential property. She wants to interactively select a parcel, find all the single-family residential lots within a specified distance, and analyze the selected features. This involves using a feature set variable to prompt the user to select a parcel, setting a variable to accept the buffer distance, selecting the residential parcels within that distance, and performing a summary statistics process.

In the exercise, Ms. Houston wants to have this model as interactive and flexible as possible, because she envisions running it many times for property all over the city as part of her overall tax value study. Tax data is available in the TaxValues feature class in the Property Data feature dataset, which is in the CityOfOleander geodatabase within the Data folder. The field TaxVal06 is the most current set of values. You'll need to use a feature set variable in your model so she can select a parcel to begin the process. Then, you'll use the Select By Location tool with the optional distance parameter to select surrounding parcels. The distance value could be collected as a stand-alone variable, or you could expose the parameter of the Select By Location tool as a variable and make it a model parameter. That selection set includes all parcels, but you'll want to perform the analysis on only the residential property. The field DU contains the number of dwelling units on each property, and those with a value of 1 are single-family homes. The Select Layer By Attribute tool selects those properties.

Finally, Ms. Houston would like to get the highest-valued home in the set, the lowest-valued home in the set, and the average home value in the set. The Summary Statistics tool can produce these results in a new table. You should prompt the user for a new output table name each time so that the files don't overwrite each other on subsequent runs of the model.

Before you take on the challenge, examine the steps needed to complete the task:

- Use a point-type feature set variable, which prompts the user to click a single point on the map.
- Set up a variable to accept the buffer distance for the Select Layer By Location tool. As a bonus, accept multiple distances and perform the analysis once for each distance provided. **Tip:** set up the variable as a stand-alone list variable and use variable substitution to include the distance in the output table name.
- Use the point from the feature set and the specified distance as input for the Select By Location tool to select property within a distance of the point.
- Use the Select Layer By Attribute tool and the REMOVE_FROM_SELECTION option to take out parcels where DU <> 1.
- Use the Summary Statistics tool to get the minimum, maximum, and mean house values of TaxVal06. Be sure to prompt the user for the name of this table. As a bonus: find the minimum, maximum, and mean house values for each subdivision in the selected set. **Tip:** set the case value in the Summary Statistics tool to Prop-Des-1 (subdivision name).

If you want to make the model run for multiple distances, you'll need to create one model to prompt the user for the subject parcel, and then create another model to perform the summary statistics. The second model is called when the first model successfully completes its run. You may even want to verify that a parcel was selected before calling the second model, in case Ms. Houston clicks on a roadway or in a neighboring city. When the model is running correctly, it should produce a summary table for each distance provided.

Challenge 5: Determining library patron density

The Oleander Library is seeking funding for a new bookmobile, and it needs a map exhibit to convince the city council that the library patrons would benefit from this type of service. In a previous analysis project, the library took the patron locations and aggregated them for each location, so that for places such as apartment complexes, there is a single point feature with an attribute showing the number of patrons in that complex. Otherwise, many patrons would have identical locations. This feature class was created from the PatronLocation layer using the Collect Events tool.

For this analysis, the library would like to show a buffered density, meaning that it would like to see where the patrons are, and add a density measurement showing how many other patrons are near them. The areas with the largest density would be the prime spots for the bookmobile to visit.

The data to be used in the exercise is in the PatronLocation_Aggregated feature class in the LibraryData feature dataset, which is in the CityOfOleander geodatabase within the Data folder. The attribute table contains a field called ICOUNT, which contains the number of patrons at a particular location. What the library wants is to select a location and count the number of patrons within 2,000 feet. This will be a summary of the ICOUNT field, not a count of features.

Before you take on the challenge, examine the steps needed to complete the task:

- Create a new field to store the results for each feature.
- Select a feature from the PatronLocation_Aggregated feature class.
- Select all the locations within 2,000 feet.
- Perform a summary on the ICOUNT field.
- Get the summary value and write it in the new field.
- Use an iterator to select the next feature and repeat the process.
- As a bonus, have the model iterate through the Oleander Library geodatabase and perform this analysis on the multiple years' worth of data stored there. **Tip:** start with the Collect Events tool.

When the model completes all the calculations, you can symbolize the data by the new field values and see the buffered patron density.

Data source credits

Exercise 1a
\ESRIPress\GTKModelbuilder\Data\CityOfOleander.mdb\Accident Information\AccidentTemplate, created by the author

\ESRIPress\GTKModelbuilder\Data\CityOfOleander.mdb\Property Data\LotBoundaries, derived from City of Euless

\ESRIPress\GTKModelbuilder\Data\CityOfOleander.mdb\Property Data\Parcels, derived from City of Euless

Exercise 1b
\ESRIPress\GTKModelbuilder\Data\CityOfOleander.mdb\Accident Information\AccidentTemplate, created by the author

\ESRIPress\GTKModelbuilder\Data\CityOfOleander.mdb\Property Data\LotBoundaries, derived from City of Euless

\ESRIPress\GTKModelbuilder\Data\CityOfOleander.mdb\Property Data\Parcels, derived from City of Euless

Exercise 1c
\ESRIPress\GTKModelbuilder\Data\CityOfOleander.mdb\Accident Information\AccidentTemplate, created by the author

\ESRIPress\GTKModelbuilder\Data\CityOfOleander.mdb\Property Data\LotBoundaries, derived from City of Euless

\ESRIPress\GTKModelbuilder\Data\CityOfOleander.mdb\Property Data\Parcels, derived from City of Euless

Exercise 1f
\ESRIPress\GTKModelbuilder\Data\CityOfOleander.mdb\Accident Information\AccidentTemplate, created by the author

\ESRIPress\GTKModelbuilder\Data\CityOfOleander.mdb\Property Data\LotBoundaries, derived from City of Euless

\ESRIPress\GTKModelbuilder\Data\CityOfOleander.mdb\Property Data\Parcels, derived from City of Euless

Exercise 2a
\ESRIPress\GTKModelbuilder\Data\CityOfOleander.mdb\Accident Information\AccidentTemplate, created by the author

\ESRIPress\GTKModelbuilder\Data\CityOfOleander.mdb\Property Data\LotBoundaries, derived from City of Euless

\ESRIPress\GTKModelbuilder\Data\CityOfOleander.mdb\Property Data\Parcels, derived from City of Euless

Exercise 2b
\ESRIPress\GTKModelbuilder\Data\CityOfOleander.mdb\Accident Information\AccidentTemplate, created by the author

\ESRIPress\GTKModelbuilder\Data\CityOfOleander.mdb\StreetData\Street_Centerlines, derived from City of Euless

\ESRIPress\GTKModelbuilder\Data\CityOfOleander.mdb\Property Data\LotBoundaries, derived from City of Euless

\ESRIPress\GTKModelbuilder\Data\CityOfOleander.mdb\Property Data\Parcels, derived from City of Euless

Exercise 2c
\ESRIPress\GTKModelbuilder\Data\CityOfOleander.mdb\StreetData\Street_Centerlines, derived from City of Euless

\ESRIPress\GTKModelbuilder\Data\CityOfOleander.mdb\Property Data\LotBoundaries, derived from City of Euless

\ESRIPress\GTKModelbuilder\Data\CityOfOleander.mdb\Property Data\Parcels, derived from City of Euless

Exercise 2d

\ESRIPress\GTKModelbuilder\Data\CityOfOleander.mdb\StreetData\Street_Centerlines, derived from City of Euless

\ESRIPress\GTKModelbuilder\Data\CityOfOleander.mdb\Property Data\LotBoundaries, derived from City of Euless

\ESRIPress\GTKModelbuilder\Data\CityOfOleander.mdb\Property Data\Parcels, derived from City of Euless

Exercise 3a

\ESRIPress\GTKModelbuilder\Data\CityOfOleander.mdb\Accident Information\AccidentTemplate, created by the author

\ESRIPress\GTKModelbuilder\Data\CityOfOleander.mdb\Property Data\LotBoundaries, derived from City of Euless

\ESRIPress\GTKModelbuilder\Data\CityOfOleander.mdb\Property Data\Parcels, derived from City of Euless

Exercise 3b

\ESRIPress\GTKModelbuilder\Data\CityOfOleander.mdb\StreetData\Street_Centerlines, derived from City of Euless

\ESRIPress\GTKModelbuilder\Data\CityOfOleander.mdb\Property Data\LotBoundaries, derived from City of Euless

\ESRIPress\GTKModelbuilder\Data\CityOfOleander.mdb\Property Data\Parcels, derived from City of Euless

Exercise 3c

\ESRIPress\GTKModelbuilder\Data\CityOfOleander.mdb\Accident Information\AccidentTemplate, created by the author

\ESRIPress\GTKModelbuilder\Data\CityOfOleander.mdb\Property Data\LotBoundaries, derived from City of Euless

\ESRIPress\GTKModelbuilder\Data\CityOfOleander.mdb\Property Data\Parcels, derived from City of Euless

\ESRIPress\GTKModelbuilder\Data\CityOfOleander.mdb\StreetData\Street_Centerlines, derived from City of Euless

Exercise 3d

\ESRIPress\GTKModelbuilder\Data\CityOfOleander.mdb\Accident Information\AccidentTemplate, created by the author

\ESRIPress\GTKModelbuilder\Data\CityOfOleander.mdb\Property Data\LotBoundaries, derived from City of Euless

\ESRIPress\GTKModelbuilder\Data\CityOfOleander.mdb\Property Data\Parcels, derived from City of Euless

\ESRIPress\GTKModelbuilder\Data\CityOfOleander.mdb\StreetData\Street_Centerlines, derived from City of Euless

Exercise 3e

\ESRIPress\GTKModelbuilder\Data\CityOfOleander.mdb\Accident Information\AccidentTemplate, created by the author

\ESRIPress\GTKModelbuilder\Data\CityOfOleander.mdb\Property Data\LotBoundaries, derived from City of Euless

\ESRIPress\GTKModelbuilder\Data\CityOfOleander.mdb\Property Data\Parcels, derived from City of Euless

\ESRIPress\GTKModelbuilder\Data\CityOfOleander.mdb\StreetData\Street_Centerlines, derived from City of Euless

Exercise 4a

\ESRIPress\GTKModelbuilder\Data\CityOfOleander.mdb\Property Data\Parcels, derived from City of Euless

Exercise 4b

\ESRIPress\GTKModelbuilder\Data\CityOfOleander.mdb\Property Data\Parcels, derived from City of Euless

Exercise 4c

\ESRIPress\GTKModelbuilder\Data\CityOfOleander.mdb\Property Data\Parcels, derived from City of Euless

\ESRIPress\GTKModelbuilder\Data\CityOfOleander.mdb\Property Data\CityArea, derived from City of Euless

\ESRIPress\GTKModelbuilder\Data\OleanderLibrary.mdb\Year 2000\PatronActivity2000, derived from City of Euless

Exercise 5a

\ESRIPress\GTKModelbuilder\Data\CityOfOleander.mdb\Property Data\CityArea, derived from City of Euless

\ESRIPress\GTKModelbuilder\Data\RFDA Data.mdb\Base Data\RFDA_Streets_NCTCOG, courtesy of NCTCOG

Exercise 5b

\ESRIPress\GTKModelbuilder\Data\CityOfOleander.mdb\Property Data\CityArea, derived from City of Euless

\ESRIPress\GTKModelbuilder\Data\RFDA Data.mdb\Base Data\RFDA_Streets_NCTCOG, courtesy of NCTCOG

\ESRIPress\GTKModelbuilder\Data\OleanderLibrary.mdb\Year 1996\PatronActivity1996, derived from City of Euless

\ESRIPress\GTKModelbuilder\Data\OleanderLibrary.mdb\Year 1997\PatronActivity1997, derived from City of Euless

\ESRIPress\GTKModelbuilder\Data\OleanderLibrary.mdb\Year 1998\PatronActivity1998, derived from City of Euless

\ESRIPress\GTKModelbuilder\Data\OleanderLibrary.mdb\Year 1999\PatronActivity1999, derived from City of Euless

\ESRIPress\GTKModelbuilder\Data\OleanderLibrary.mdb\Year 2000\PatronActivity2000, derived from City of Euless

\ESRIPress\GTKModelbuilder\Data\OleanderLibrary.mdb\Year 2000\PatronActivity2000_Select, derived from City of Euless

\ESRIPress\GTKModelbuilder\Data\OleanderLibrary.mdb\Year 2001\PatronActivity2001, derived from City of Euless

\ESRIPress\GTKModelbuilder\Data\OleanderLibrary.mdb\Year 2002\PatronActivity2002, derived from City of Euless

\ESRIPress\GTKModelbuilder\Data\OleanderLibrary.mdb\Year 2003\PatronActivity2003, derived from City of Euless

\ESRIPress\GTKModelbuilder\Data\OleanderLibrary.mdb\Year 2004\PatronActivity2004, derived from City of Euless

Exercise 6a

\ESRIPress\GTKModelbuilder\Data\OleanderFireDept.mdb\Base Data\Station_1, derived from City of Euless

\ESRIPress\GTKModelbuilder\Data\OleanderFireDept.mdb\Base Data\Station_2, derived from City of Euless

\ESRIPress\GTKModelbuilder\Data\OleanderFireDept.mdb\Base Data\Station_3, derived from City of Euless

\ESRIPress\GTKModelbuilder\Data\OleanderFireDept.mdb\Response Data 2006\Response2006, derived from City of Euless

\ESRIPress\GTKModelbuilder\Data\RFDA Data.mdb\Base Data\RFDA_Streets_NCTCOG, courtesy of NCTCOG

\ESRIPress\GTKModelbuilder\Data\CityOfOleander.mdb\CountyCensus, courtesy of NCTCOG

\ESRIPress\GTKModelbuilder\Data\CityOfOleander.mdb\Property Data\CityArea, derived from City of Euless

Exercise 6b

\ESRIPress\GTKModelbuilder\Data\RFDA Data.mdb\Base Data\RFDA_Streets_NCTCOG, courtesy of NCTCOG

\ESRIPress\GTKModelbuilder\Data\CityOfOleander.mdb\CountyCensus, courtesy of NCTCOG

\ESRIPress\GTKModelbuilder\Data\CityOfOleander.mdb\Property Data\CityArea, derived from City of Euless

Exercise 6c

\ESRIPress\GTKModelbuilder\Data\RFDA Data.mdb\Base Data\RFDA_Streets_NCTCOG, courtesy of NCTCOG

\ESRIPress\GTKModelbuilder\Data\CityOfOleander.mdb\CountyCensus, courtesy of NCTCOG

\ESRIPress\GTKModelbuilder\Data\CityOfOleander.mdb\Property Data\CityArea, derived from City of Euless

Exercise 6d

\ESRIPress\GTKModelbuilder\Data\RFDA Data.mdb\Base Data\RFDA_Streets_NCTCOG, courtesy of NCTCOG

\ESRIPress\GTKModelbuilder\Data\RFDA Data.mdb\Base Data\RFDAStations, derived from City of Euless

Exercise 6e

\ESRIPress\GTKModelbuilder\Data\OleanderContest.mdb\Year2008\PatronActivity2008, derived from City of Euless

\ESRIPress\GTKModelbuilder\Data\OleanderContest.mdb\Year2009\PatronActivity2009, derived from City of Euless

\ESRIPress\GTKModelbuilder\Data\OleanderContest.mdb\Year2009\PatronActivity2009Test, derived from City of Euless

\ESRIPress\GTKModelbuilder\Data\OleanderContest.mdb\Year2010\PatronActivity2010, derived from City of Euless

Exercise 6f

\ESRIPress\GTKModelbuilder\Data\RFDA Data.mdb\Base Data\RFDA_Streets_NCTCOG, courtesy of NCTCOG

\ESRIPress\GTKModelbuilder\Data\OleanderLibrary.mdb\LibraryData\ApartmentComplexes, derived from City of Euless

\ESRIPress\GTKModelbuilder\Data\OleanderLibrary.mdb\Year 2004\PatronActivity2004, derived from City of Euless

Exercise 7a

\ESRIPress\GTKModelbuilder\Data\RFDA Shapefiles\BedfordCalls.shp, courtesy of City of Bedford

\ESRIPress\GTKModelbuilder\Data\RFDA Shapefiles\ColleyvilleCalls.shp, courtesy of City of Colleyville

\ESRIPress\GTKModelbuilder\Data\RFDA Shapefiles\HurstCalls.shp, courtesy of City of Hurst

\ESRIPress\GTKModelbuilder\Data\RFDA Shapefiles\KellerCalls.shp, courtesy of City of Keller

\ESRIPress\GTKModelbuilder\Data\RFDA Shapefiles\OleanderCalls.shp, courtesy of City of Euless

\ESRIPress\GTKModelbuilder\Data\RFDA Shapefiles\SouthlakeCalls.shp, courtesy of City of Southlake

Appendix A challenges

\ESRIPress\GTKModelbuilder\SampleModels\Public Library.gdb\PatronActivity2010, derived from City of Euless

Data license agreement

Important: Read carefully before opening the sealed media package

Environmental Systems Research Institute Inc. (Esri) is willing to license the enclosed data and related materials to you only upon the condition that you accept all of the terms and conditions contained in this license agreement. Please read the terms and conditions carefully before opening the sealed media package. By opening the sealed media package, you are indicating your acceptance of the Esri License Agreement. If you do not agree to the terms and conditions as stated, then Esri is unwilling to license the data and related materials to you. In such event, you should return the media package with the seal unbroken and all other components to Esri.

Esri license agreement

This is a license agreement, and not an agreement for sale, between you (Licensee) and Environmental Systems Research Institute Inc. (Esri). This Esri License Agreement (Agreement) gives Licensee certain limited rights to use the data and related materials (Data and Related Materials). All rights not specifically granted in this Agreement are reserved to Esri and its Licensors.

Reservation of Ownership and Grant of License: Esri and its Licensors retain exclusive rights, title, and ownership to the copy of the Data and Related Materials licensed under this Agreement and, hereby, grant to Licensee a personal, nonexclusive, nontransferable, royalty-free, worldwide license to use the Data and Related Materials based on the terms and conditions of this Agreement. Licensee agrees to use reasonable effort to protect the Data and Related Materials from unauthorized use, reproduction, distribution, or publication.

Proprietary Rights and Copyright: Licensee acknowledges that the Data and Related Materials are proprietary and confidential property of Esri and its Licensors and are protected by United States copyright laws and applicable international copyright treaties and/or conventions.

Permitted Uses: Licensee may install the Data and Related Materials onto permanent storage device(s) for Licensee's own internal use.

Licensee may make only one (1) copy of the original Data and Related Materials for archival purposes during the term of this Agreement unless the right to make additional copies is granted to Licensee in writing by Esri.

Licensee may internally use the Data and Related Materials provided by Esri for the stated purpose of GIS training and education.

Uses Not Permitted: Licensee shall not sell, rent, lease, sublicense, lend, assign, time-share, or transfer, in whole or in part, or provide unlicensed Third Parties access to the Data and Related Materials or portions of the Data and Related Materials, any updates, or Licensee's rights under this Agreement.

Licensee shall not remove or obscure any copyright or trademark notices of Esri or its Licensors.

Term and Termination: The license granted to Licensee by this Agreement shall commence upon the acceptance of this Agreement and shall continue until such time that Licensee elects in writing to discontinue use of the Data or Related Materials and terminates this Agreement. The Agreement shall automatically terminate without notice if Licensee fails to comply with any provision of this Agreement. Licensee shall then return to Esri the Data and Related Materials. The parties hereby agree that all provisions that operate to protect the rights of Esri and its Licensors shall remain in force should breach occur.

Disclaimer of Warranty: The Data and Related Materials contained herein are provided "as-is," without warranty of any kind, either express or implied, including, but not limited to, the implied warranties of merchantability, fitness for a particular purpose, or noninfringement. Esri does not warrant that the Data and Related Materials will meet Licensee's needs or expectations, that the use of the Data and Related Materials will be uninterrupted, or that all nonconformities, defects, or errors can or will be corrected. Esri is not inviting reliance on the Data or Related Materials for commercial planning or analysis purposes, and Licensee should always check actual data.

Data Disclaimer: The Data used herein has been derived from actual spatial or tabular information. In some cases, Esri has manipulated and applied certain assumptions, analyses, and opinions to the Data solely for educational training purposes. Assumptions, analyses, opinions applied, and actual outcomes may vary. Again, Esri is not inviting reliance on this Data, and the Licensee should always verify actual Data and exercise their own professional judgment when interpreting any outcomes.

Limitation of Liability: Esri shall not be liable for direct, indirect, special, incidental, or consequential damages related to Licensee's use of the Data and Related Materials, even if Esri is advised of the possibility of such damage.

No Implied Waivers: No failure or delay by Esri or its Licensors in enforcing any right or remedy under this Agreement shall be construed as a waiver of any future or other exercise of such right or remedy by Esri or its Licensors.

Order for Precedence: Any conflict between the terms of this Agreement and any FAR, DFAR, purchase order, or other terms shall be resolved in favor of the terms expressed in this Agreement, subject to the government's minimum rights unless agreed otherwise.

Export Regulation: Licensee acknowledges that this Agreement and the performance thereof are subject to compliance with any and all applicable United States laws, regulations, or orders relating to the export of data thereto. Licensee agrees to comply with all laws, regulations, and orders of the United States in regard to any export of such technical data.

Severability: If any provision(s) of this Agreement shall be held to be invalid, illegal, or unenforceable by a court or other tribunal of competent jurisdiction, the validity, legality, and enforceability of the remaining provisions shall not in any way be affected or impaired thereby.

Governing Law: This Agreement, entered into in the County of San Bernardino, shall be construed and enforced in accordance with and be governed by the laws of the United States of America and the State of California without reference to conflict of laws principles. The parties hereby consent to the personal jurisdiction of the courts of this county and waive their rights to change venue.

Entire Agreement: The parties agree that this Agreement constitutes the sole and entire agreement of the parties as to the matter set forth herein and supersedes any previous agreements, understandings, and arrangements between the parties relating hereto.

Installing the data

Getting to Know ArcGIS ModelBuilder includes a DVD containing exercise data and all components necessary to complete the optional model challenges. You must already have a licensed copy of ArcGIS Desktop 10 installed on your computer (or have access to the software through a network). Use your licensed software to perform the exercises in this book.

Installing the exercise data

Follow the steps below to install the exercise data.

1 Put the data DVD in your computer's DVD drive. A splash screen will appear.

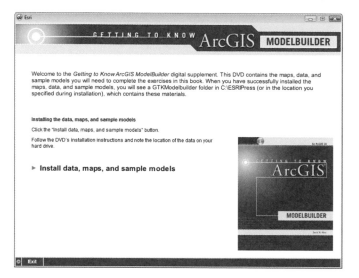

2 Read the welcome screen, and then click the Install exercise data link. This launches the InstallSheild Wizard.

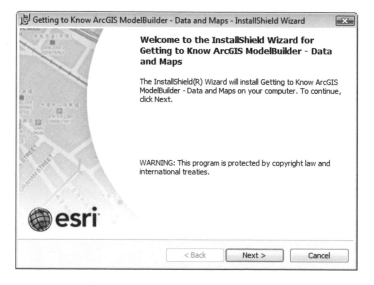

3 Click Next. Read and accept the license agreement terms, and then click next.

4 Accept the default installation folder or click Browse and navigate to the drive or folder location where you want to install the data. If you choose an alternate location, please make note of it as the book's exercises direct you to **C:\ESRIPress\GTKModelbuilder.**

5 Click Next. The installation will take some time. When the installation is complete, you see the following message:

6 Click Finish. The exercise data is installed on your computer in a folder called GTKModelbuilder.

If you have a licensed copy of ArcGIS Desktop 10 installed on your computer, you are ready to start *Getting to Know ArcGIS ModelBuilder*.

Uninstalling the exercise data

To uninstall the exercise data from your computer, open your operating system's control panel and double-click the Add/Remove Programs icon. In the Add/Remove Programs dialog box, select the following entry and follow the prompts to remove it:

- Getting to Know ArcGIS ModelBuilder - Data and Maps

INDEX

Related titles from Esri Press

Getting to Know ArcGIS Desktop, second edition

ISBN: 978-1-58948-260-9

Getting to Know ArcGIS Desktop introduces principles of GIS as it acquaints readers with the building blocks of ArcGIS Desktop, including ArcMap for displaying and querying maps, ArcCatalog for organizing geographic data, and ModelBuilder for diagramming and processing solutions to complex spatial analysis problems.

GIS, Spatial Analysis, and Modeling

ISBN: 978-1-58948-130-5

GIS, Spatial Analysis, and Modeling presents a compendium of papers discussing the state of the art in computerized spatial analysis and modeling. Recent advancements in GIS software, coupled with the availability of spatially referenced data, now make possible the sophisticated modeling and statistical analysis of all types of geographic phenomena. This book serves as a resource for geographic analysts, modelers, software engineers, and GIS professionals.

Smart Land-Use Analysis: The LUCIS Model

ISBN: 978-1-87910-174-9

Smart Land-Use Analysis: The LUCIS Model examines the land-use conflict identification strategy (LUCIS), applied with great success by a nine-county region of north central Florida. The LUCIS model uses the ArcGIS geoprocessing framework, particularly ModelBuilder, to analyze suitability and preference for major land-use categories, determine potential future conflict, and build future land-use scenarios.

GIS Tutorial 3: Advanced Workbook

ISBN: 978-1-58948-207-4

GIS Tutorial 3 features exercises that demonstrate the advanced functionality of the ArcEditor and ArcInfo licenses of ArcGIS Desktop, including the use of ModelBuilder. This workbook is divided into four sections: geodatabase framework design, data creation and management, workflow optimization, and labeling and symbolizing.

Esri Press publishes books about the science, application, and technology of GIS. Ask for these titles at your local bookstore or order by calling 1-800-447-9778. You can also read book descriptions, read reviews, and shop online at www.esri.com/esripress. Outside the United States, visit our Web site at www.esri.com/esripressorders for a full list of book distributors and their territories.